Culture + Leisure Services
Red Doles Lane
Huddersfield, West Yorks. HD2 1YF

Einstein's Miraculous Year

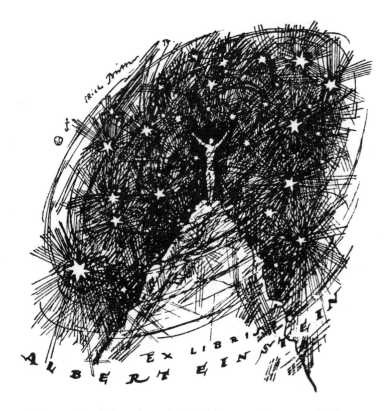

Einstein's bookplate, drawn by Erich Büttner. (Courtesy of Hebrew
University of Jerusalem)

Einstein's Miraculous Year

✦

*Five Papers That Changed
the Face of Physics*

EDITED AND WITH A NEW INTRODUCTION BY
John Stachel

WITH THE ASSISTANCE OF
Trevor Lipscombe, Alice Calaprice,
and Sam Elworthy

AND

WITH A FOREWORD BY
Roger Penrose

PRINCETON UNIVERSITY PRESS
PRINCETON AND OXFORD

The Library of Congress has cataloged the cloth edition of this book as follows
Einstein's miraculous year : five papers that changed the
face of physics / edited and introduced by John Stachel ;
with the assistance of Trevor Lipscombe,
Alice Calaprice, Sam Elworthy ;
with a foreword by Roger Penrose.
p.   cm.
ISBN 0-691-05938-1 (cloth : alk. paper)
1. Physics—History—20th century.   2. Einstein, Albert,
1879–1955.   I. Stachel, John J., 1928–   .
QC7.E52   1998
530.1—dc21   97-48441

British Library Cataloging-in-Publication Data is available

This book has been composed in Caledonia

Printed on acid-free paper. ∞

pup.princeton.edu

Printed in the United States of America

4   6   8   10   9   7   5   3

CONTENTS

FOREWORD

In the twentieth century, we have been greatly privileged to witness *two* major revolutions in our physical picture of the world. The first of these upturned our conceptions of space and time, combining the two into what we now call *space-time*, a space-time which is found to be subtly curved in a way that gives rise to that long-familiar, omnipresent but mysterious, phenomenon of gravity. The second of these revolutions completely changed the way in which we understand the nature of matter and radiation, giving us a picture of reality in which particles behave like waves and waves like particles, where our normal physical descriptions become subject to essential uncertainties, and where individual objects can manifest themselves in several places at the same time. We have come to use the term "relativity" to encompass the first of these revolutions and "quantum theory" to encompass the second. Both have now been observationally confirmed to a precision unprecedented in scientific history.

I think that it is fair to say that there are only three previous revolutions in our understanding of the physical world that can bear genuine comparison with either. For the first of these three, we must turn back to ancient Greek times, where the notion of Euclidean geometry was introduced and some conception was obtained of rigid bodies and static configurations. Moreover, there was a beginning of an appreciation of the crucial role of *mathematical reasoning* in our insights into Nature. For the second of the three, we must leap to the seventeenth century, when Galileo and Newton told us how the motions of ponderable bodies can be understood in terms of forces between their constituent particles and the accelerations that these forces engender. The nineteenth century gave us the third revolution, when Faraday and Maxwell showed us that particles were not enough, and we must consider, also, that there are continuous fields pervading space, with a reality as great as that of the particles themselves. These fields were combined into a single all-pervasive entity, referred to as the *electromagnetic field*, and the behavior of light could be beautifully explained in terms of its self-propagating oscillations.

Turning now to our present century, it is particularly remarkable that a single physicist—Albert Einstein—had such extraordinarily deep perceptions of the workings of Nature that he laid foundation stones of *both* of these twentieth-century revolutions in the *single year* of 1905. Not only that, but in this same year Einstein also provided fundamental new insights into two other areas, with his doctoral dissertation on the determination of molecular dimensions and with his analysis of the nature of Brownian motion. This latter analysis alone would have earned Einstein a place in history. Indeed, his work on Brownian motion (together with the independent and parallel work of Smoluchowski) laid the foundations of an important piece of statistical understanding which has had enormous implications in numerous other fields.

This volume brings together the five papers that Einstein published in that extraordinary year. To begin with, there is the one just referred to on molecular dimensions (paper 1), followed by the one on Brownian motion (paper 2). Then come two on the special theory of relativity: the first initiates the "relativity" revolution, now so familiar to physicists (and also perceived by the public at large), in which the notion of absolute time is abolished (paper 3); the second is a short note deriving Einstein's famous "$E = mc^2$" (paper 4). Finally, the (only) paper that Einstein himself actually referred to as "revolutionary" is presented, which argues that we must, in some sense, return to the (Newtonian) idea that light consists of *particles* after all—just when we had become used to the idea that light consists solely of electromagnetic waves (paper 5). From this apparent paradox, an important ingredient of quantum mechanics was born. Together with these five classic Einstein papers, John Stachel has provided fascinating and highly illuminating introductions that set Einstein's achievements in their appropriate historical settings.

I have referred above to the twentieth century's two extraordinary revolutions in physical understanding. But it should be made clear that, fundamental as they were, Einstein's papers of 1905 did not quite provide the initial shots of those revolutions; nor did these particular papers set out the final nature of their new regimes.

The revolution in our picture of space and time that Einstein's two 1905 relativity papers provided concerned only what we now call the *special* theory. The full formulation of the *general* theory of relativity, in which gravitation is interpreted in terms of curved space-time geometry, was not achieved until ten years later. And even for special relativity, the wonderful insights presented by Einstein in 1905 provided a theory that was not totally original with him, this theory having been grounded in earlier ideas (notably those of Lorentz and Poincaré). Moreover, Einstein's viewpoint in 1905 still lacked one important further insight—that of *space-time*—introduced by Hermann Minkowski three years later. Minkowski's notion of a four-dimensional space-time was soon adopted by Einstein, and it became one of the crucial steppingstones to what was later to become Einstein's crowning achievement: his *general* theory of relativity. With regard to quantum mechanics, the initial shots of this revolution had been Max Planck's extraordinary papers of 1900, in which the famous relation $E = h\nu$ was introduced, asserting that energy of radiation is produced in discrete little bundles, in direct proportion to the radiation's frequency. But Planck's ideas were hard to make sense of in terms of the ordinary physics of the day, and only Einstein seems to have realized (after some while) that these tentative proposals had a fundamental significance. Quantum theory itself took many years to find its appropriate formulation— and this time the unifying ideas came not from Einstein, but from a number of other physicists, most notably Bohr, Heisenberg, Schrödinger, Dirac, and Feynman.

There are some remarkable aspects to Einstein's relation to quantum physics, which border almost on the paradoxical. Earliest and perhaps most striking of these seeming paradoxes is the fact that Einstein's initial revolutionary papers on quantum phenomena (paper 5) and on relativity (paper 3) appear to start from mutually contradictory standpoints with regard to the status of Maxwell's electromagnetic theory as an explanation of light. In paper 5, Einstein explicitly rejects the view that Maxwell's equations suffice to explain the actions of light (as waves in the electromagnetic field) and he puts forward a model in which light behaves, instead, like little particles. Yet, in (the later) paper 3, he develops the special theory of relativity from the starting point that Maxwell's theory indeed *does* represent fundamental truth, and the relativity theory that Einstein constructs is specifically designed so that it leaves Maxwell's equations intact. Even at the beginning of paper 5 itself, where Einstein puts forward a "particle" viewpoint of light in conflict with Maxwell's theory, he comments on the latter (wave) theory of light that it "will probably never be replaced by another theory." This seeming conflict is all the more striking when one considers that Einstein's incredible strength as a physicist came from his direct physical insights into the workings of Nature. One could well imagine some lesser figure "trying out" one model and then another (as is common practice with physicists of today), where the contradiction between the two proposed viewpoints would cause no real concern, since neither carries any particular conviction. But with Einstein, things were quite different. He appears to have had very clear and profound ideas as to what Nature was "really like" at levels not readily perceivable by other physicists. Indeed, his ability to perceive Nature's reality was one of his particular strengths. To me, it is virtually inconceivable that he would have put forward two papers in the same year which depended upon hypothetical views of Nature that he felt were in contradiction with each other. Instead, he must have felt (correctly, as it turned out) that "deep down" there was no real contradiction between the accuracy—indeed "truth"—of Maxwell's wave theory and the alternative "quantum" particle view that he put forward in paper 5.

One is reminded of Isaac Newton's struggles with basically the same problem—some 300 years earlier—in which he proposed a curious hybrid of a wave and particle viewpoint in order to explain conflicting aspects of the behavior of light. In Newton's case, it is possible to understand his dogged adherence to a particle-type picture if one takes the (reasonable) view that Newton wished to preserve a relativity principle. But this argument holds only if the relevant relativity principle is that of Galileo (and Newton). In Einstein's case, such an argument will not do, for the reason that he explicitly put forward a *different* relativity principle from the Galilean one, in which Maxwell's wave theory could survive intact. Thus, it is necessary to look more deeply to find the profound reasons for Einstein's extraordinary conviction that although Maxwell's wave picture of light was, in some sense, "true"—having been well established in 1905—it nevertheless needed to be altered to something different which, in certain respects, harked back Newton's hybrid "wave-particle" picture of three centuries earlier.

It would seem that one of the important influences that guided Einstein was his awareness of the conflict between the discrete nature of the particles constituting ponderable bodies and the continuous nature of Maxwell's fields. It is particularly manifest in Einstein's 1905 papers that this conflict was very much in his mind. In papers 1 and 2, he was directly concerned with demonstrating the nature of the molecules and other small particles which constitute a fluid, so the "atomic" nature of matter was indeed at the forefront. In these papers, he showed himself to be a master of the physical/statistical techniques required. In paper 5, he put this extraordinary expertise to use by treating electromagnetic fields in the same way, thereby explaining effects that cannot be obtained with the Maxwellian view of light alone. Indeed, it was made clear by Einstein that the problem with the classical approach was that a picture in which continuous fields and discrete particles *coexist*, each interacting with the other, does not really make physical sense. Thus, he initiated an important step toward the present-day quantum-theoretic viewpoint that particles must indeed take on attributes of waves, and fields must take on attributes of particles. Looked at appropriately in the quantum picture, particles and waves actually turn out to be the same thing.

The question is often raised of another seeming paradox: Why, when Einstein started from a vantage point so much in the lead of his contemporaries with regard to understanding quantum phenomena, was he nevertheless left behind by them in the subsequent development of quantum theory? Indeed, Einstein never even accepted the quantum theory, as that theory finally emerged in the 1920s. Many would hold that Einstein was hampered by his "outdated" *realist* standpoint, whereas Niels Bohr, in particular, was able to move forward simply by denying the very existence of such a thing as "physical reality" at the quantum level of molecules, atoms, and elementary particles. Yet, it is clear that the fundamental advances that Einstein was able to achieve in 1905 depended crucially on his robust adherence to a belief in the *actual* reality of physical entities at the molecular and sub-molecular levels. This much is particularly evident in the five papers presented here.

Can it really be true that Einstein, in any significant sense, was as profoundly "wrong" as the followers of Bohr might maintain? I do not believe so. I would, myself, side strongly with Einstein in his belief in a submicroscopic reality, and with his conviction that present-day quantum mechanics is fundamentally incomplete. I am also of the opinion that there are crucial insights to be found as to the nature of this reality that will ultimately come to light from a profound analysis of a seeming conflict between the underlying principles of quantum theory and those of Einstein's own general relativity. It seems to me that only when such insights are at hand and put appropriately to use will the fundamen-

INTRODUCTION
TO THE CENTENARY EDITION

Long live impudence! It is my guardian angel in this world.
Einstein to Mileva Marić, 12 December 1901[1]

I seek solitude, only then silently to lament it.
Einstein to "Mama" Winteler, 21 May 1897[2]

A full century—Einstein's century, according to *Time* magazine[3]—has elapsed since "Einstein's miraculous year." *Time's* cover featuring Einstein as the aged sage and saint of legend symbolizes the formidable barrier between us and the defiant yet vulnerable twenty-six-year-old of 1905.[4] And I fear that most of the planned centennial celebrations may continue to foster the myth of an Einstein born ancient, at best looking at the young Einstein through the distorting lens of the old.

So let us try to look directly at the young Einstein, starting with the child and stopping with the Swiss Patent Office clerk of 1905. The original introduction tried to explain the nature and significance of Einstein's scientific achievements in that year.[5] Here I shall discuss some elements of his familial background and personal characteristics that helped to shape the young Einstein.

I have organized this highly selective survey around four themes:

1. The role of some polarities in the development of the young Einstein's character.

I thank Alberto Martinez and Karen Jones for their careful reading of an earlier draft and their many helpful suggestions for its improvement.

2. The technological milieu in which he grew up and its effect on his development.
3. The nature of his thought process as he described it.
4. His failed attempt to combine work and intimacy.

1. SOME POLARITIES OF THE YOUNG EINSTEIN'S CHARACTER

Einstein's behavior provides evidence of the youthful conflicts in his personality between a number of complementary but opposing trends, which I shall refer to as polarities.[6] I shall focus on two such polarities:

1. the striving for recognition and approval from elders in positions of authority, and the need to maintain independence, and sometimes even to show defiance of such authority figures ("Long live impudence!"), in order to pursue his own goals; and
2. the urge for close companionship and intimate relationships, and the need for solitude ("I seek solitude . . .") in order to pursue his intellectual "inventions" (see section 3).

In a short but profound study, after reviewing the evidence on Einstein's early childhood, the famed psychoanalyst Erik Erikson asks rhetorically:

Were the boy's symptoms [beginning with his comparatively late start in speech, discussed below in section 3] due to an outright *defect* or to a systematic *difference* in development; or were they also reinforced by a mighty *diffidence*—or, eventually, even some *defiance*?[7]

Erikson continues:

[L]ittle Albert had it in him to object to having to learn anything in any but his own way. In his early childhood this could be expressed in a sudden rage (against a private teacher, for example) that was attributed to a disposition inherited from his maternal grandfather. Later, the resistance against enforced instruction, far from ever being "broken," became a deep and basic character trait that permitted the child and the youth to remain free in learning, no matter how slowly or by what sensory or cognitive steps he accomplished it.[8]

Einstein was thus able to defend himself internally against the considerable pressures to learn by rote and other conventional methods, and to resist devoting extra effort to subjects, such as foreign languages, in which he was weak or uninterested. This is quite a different thing from saying that he was a poor student. He was an excellent learner in the subjects on which he chose to concentrate, developing habits of independent study that took him far beyond his peers in such areas as mathematics, physics, and philosophy.

He started his schooling Munich, at a Catholic primary school,[9] where he received good grades; but his experiences there served to set up a barrier between him and the other pupils. The only Jewish pupil in his class, he later recalled:

> Among the children, particularly in elementary school, anti-Semitism was rife. It was based on racial characteristics that, remarkably enough, were known to the children, and on impressions from religious education. Actual attacks and insults on the way to school were frequent, but generally not too bad. At any rate, they sufficed to instill in the child a strong feeling of alienation.[10]

When he was nine, his parents sent him to a prestigious new academic secondary school, the Luitpold Gymnasium. Their choice of a *Gymnasium*, with its emphasis on classical languages and literature, was somewhat unusual. It was more common among well-to-do Jewish families such as the Einsteins (see section 2 below) to send their sons to *Realschule*, where the education centered around modern culture, the sciences, and technology. In retrospect, Einstein apparently felt this would have been the better choice, later writing to his son Hans Albert:

> I am quite in agreement that you go into the *Realgymnasium*. For a person whose talents lie in the direction that yours do, it is not good if he is crammed all too full with languages.[11]

Einstein's only sibling, his sister Maja, describes his Gymnasium years:

> [H]e felt very discontented at the *Gymnasium*. The style of teaching in most subjects was repugnant to him, and in addition the teachers seemed not to be very favorably inclined to him.[12] Also particularly distasteful to the youngster was the military tone of the school, the systematic upbringing to worship of authority, which was already supposed to accustom the pupils to military discipline. With horror he thought of the not-too-distant time, when he would have to put on a soldier's uniform in order to satisfy his military service obligations. Oppressed mentally and having become nervous he looked for a way out.[13]

To the consternation of his parents, who had left him behind to finish school when they moved to Italy in 1895 in search of better business opportunities (see section 2 below), the sixteen-year-old Einstein abruptly left the Gymnasium in the middle of the school year and joined his family. Although he was only sixteen and still lacked a year-and-a-half of secondary schooling, he applied for entrance to the Swiss Federal Polytechnic in Zurich and was allowed to take the entrance examination. Though his grades in mathematics and physics impressed H. F. Weber, professor of physics at the Poly, he was advised to finish his secondary education at the nearby Aargau Kantonsschule, graduation from which automatically ensured admission to the Poly.[14] There he found a quite different educational milieu, in which he started to flourish. Hans Byland, a fellow student, later recalled:

> At the Aargau Cantonal School in the 1890s a strong wind of skepticism prevailed, as is already indicated by the fact that neither from my class nor in the next two did any theologian emerge. The impudent Swabian [Einstein] didn't fit badly into this atmosphere.[15]

The choice of subjects and especially the less rigid style of instruction and more informal relation between teachers and students (in the Pestalozzi tradition) suited him. But his defiant streak persisted. When his geology teacher, Professor Fritz Mühlberg, of whom Einstein was actually quite fond, asked him on a geology expedition, "Now, Einstein, how do the strata run here, from below to above or from above to below?" Einstein cheekily replied, "It's all the same to me, Professor."[16]

Another fellow student left the following portrait:

> Unhampered by convention, his attitude towards the world was that of a laughing philosopher and his witty mockery pitilessly lashed any conceit or pose. In conversation he always had something to give. His well-schooled taste acquired from traveling—his parents lived at Milan—gave him a maturity of judgement. He made no bones about voicing his personal opinions whether they offended or not. This courageous love of truth gave his whole personality a certain cachet which, in the long run, was bound to impress even his opponents.[17]

After graduating from the Aargau School with generally high grades, except in French, he attended the Swiss Federal Polytechnical Institute in Zurich. Again, the defiant streak surfaced from time to time. Another Poly student, Margarete von Uexküll, took the same experimental physics course as Einstein:

> She had spent the whole of a warm June afternoon wrestling with an experiment in the Polytechnic's laboratory. Frustration overwhelming her, she was drawn into an argument with a small, fat physics Professor [Jean Pernet], who refused to let her seal a test-tube with a cork for fear it would break. Suddenly she noticed "a pair of the most extraordinarily large shining eyes that were clearly warning me." These belonged to Einstein, who quietly assured her that the professor was mad and had recently fainted during an angry fit in front of his class. He suggested that she give him her laboratory notes so that he could cook up some better results. At the next review, the professor exclaimed, "There, you see. With a little goodwill, and despite my impossible methods, you can apparently work out something useful."[18]

According to von Uexküll, Einstein performed similar services for eight other students during the winter semester of 1898–1899. Evidently, Pernet was aware of Einstein's attitude toward him, if not perhaps of these actions. He gave Einstein the lowest possible grade and entered the only sanction in his Polytechnic record: "March 1899: reprimand from the Director on account of lack of diligence in the Physics Practicum."[19]

Einstein originally had good relations with the senior physics professor at the Poly, H. F. Weber, whose lectures he admired,[20] and spent most of his school time during his last two years at the Poly in Weber's laboratory, "fascinated by the direct contact with experiment" (see below). But his independent streak seems ultimately to have alienated Weber too, who is said to have told him, "You're a clever boy, Einstein, an extremely clever boy, but you have one great fault: you'll never let yourself be told anything."[21]

Among his fellow students in the small entering class of Section VIA, devoted to training teachers of mathematics and physics, he soon became close to Mileva Marić, the only woman in the group.

> She and Einstein found a common interest in their passion for the study of the great physicists, and they spent a great deal of time together. For Einstein it had always been pleasant to think in society, or more correctly, to clarify his thoughts by speaking about them [see section 3 below]. Even though Mileva Maritsch [Marić] was taciturn and responded very little, in his enthusiasm Einstein hardly noticed this.[22]

His acerbity soon alienated several of Marić's fellow-Serbian girlfriends:

> The girls seem to be cross with me, although I cannot imagine why; perhaps they are expecting me to atone for other people's sins. . . . Today Mr. Einstein made up a mocking little poem at their expense, very good but also very mean, and he is intending to give it to them. That will be really wonderful. [23]

Einstein was friendly with all the other members of his class, but became especially close to only one: Marcel Grossmann, whose father later helped him to get his first permanent job at the Swiss Patent Office. While in Zurich he also befriended Michele Besso, who remained a lifelong friend and confidant. Much later, Einstein himself reminisced about his days at the Poly:

1896–1900: Study in the Section for Specialized Teachers [of Mathematics and Physics] of the Federal Polytechnic. I soon perceived that I had to be satisfied with being a mediocre student. In order to be a good student, one had to have a facility of intellectual comprehension; zeal to concentrate one's powers on all that one was presented with; [and] a love of order to record in writing what was offered in the lectures and then to work it through conscientiously. I completely lacked all of these characteristics, as I ascertained with regret.[24]

I think here we have an example of Einstein himself reading the young man in the light of the old. At least during his first two years at the Poly, Einstein seems to have been a reasonably diligent student. He preserved carefully "worked-through" lecture notes on his physics course with H. F. Weber, taken during his second year at the Poly,[25] and wrote to Marić:

Weber lectured on heat (temperature, quantities of heat, thermal motion, dynamical theory of gases) with great mastery. I look forward from one lecture of his to the next.[26]

His notes were so good that Marić used them to study for her intermediate examinations.[27]

The Poly students took only two sets of examinations: intermediate after two years, and final after four. In the intermediate exams, Einstein scored the highest of all his cohort of five in section VIA, earning a 5.7 out of a possible 6.0. He evidently took these examinations quite seriously, later describing to Marić how he had studied together with Marcel Grossmann:

When taking such an examination, one feels so responsible for everything one thinks and does that it's as if one were in a penal institution. Isn't that so? I laughed so much about these things with Grossmann at the time—but: "Laughing on the outside, crying on the inside," as one might say.[28]

It was only after passing these exams that he started to slack off in regular course work, relying on the study of Marcel Grossmann's careful notes a few months before his final examinations and accepting "the resulting guilty conscience as by far the lesser evil."[29]

> I worked most of the time in the physics laboratory, fascinated by the direct contact with experience. The balance of the time I used, in the main, in order to study at home the works of Kirchhoff, Helmholtz, Hertz, etc.[30]

His reminiscences are confirmed by his contemporary letters (see *Collected Papers*, vol. 1). In addition to these three, the letters mention study of Boltzmann, Drude, Planck, Ostwald, and Mach.

When he wrote his first paper in 1900, he sent copies to Boltzmann and Ostwald, and was evidently eager for responses that never came. His efforts to find a position as an *Assistent* in a physics department after graduation from the Poly in 1900 similarly came to naught. A touching letter from his father to Wilhelm Ostwald depicts his state of mind in 1901:

> My son feels himself deeply unhappy now in his present joblessness and every day the idea takes firmer root in him that his career has jumped the rails and will never get back on track. . . . Since my son now honors and admires you the most, highly honored Herr Professor, among all the active scholars in physics, I allow myself to turn to you with the polite request to read the paper he published in the Annalen der Physik and perhaps send him a line or two of encouragement in order that he may regain his happiness in life and creative work.[31]

No answer was forthcoming to either father or son.

When Einstein found what he thought were errors in some work by Paul Drude, work that he otherwise valued highly, he wrote Drude eagerly, hoping to establish contact with someone he admired, and perhaps obtain his help in getting a job. But Drude's reply dashed his hopes:

> I just got home . . . and found this letter from Drude. It is such manifest proof of the wretchedness of its author that no further comment by me is necessary. From now on I'll no longer turn to such people, and will instead attack them mercilessly in the journals, as they deserve. No wonder little by little one becomes a misanthrope.[32]

One could go on citing further examples, but we can see that Einstein was much less independent and more vulnerable than he later liked to depict (and probably think of) himself. Or better put, we see evidence of the first polarity mentioned at the beginning of this section: between the need to show independence (and sometimes defiance) of authority figures, and a longing for recognition and a vulnerability to slights by such figures.

Erik Erikson has discussed the second of these polarities:

> But it behooves a psychoanalyst also to return, if ever so briefly, to what Einstein himself describes and a variety of observers confirm as the price of his creativity, namely, a certain sense of isolation. [Philipp] Frank unhesitatingly describes Einstein as a "lonely person among students, colleagues, friends, family," and he comes to the sweeping conclusion that "neither professional activity nor a family had a great significance" for him.[33] If Frank intended to include marital intimacy, then Einstein in some of his most touching letters . . . seems to have sadly agreed. . . . Yet, I must say that when one reads some of his letters and sees his reflection in some of the

surviving members of his inner circle [in 1979, Erikson met Helen Dukas and Margot Einstein], one cannot doubt that this man knew, and knew how to give, some intense sense of intimacy. A certain alternation of isolation and outgoingness (especially toward children) seems to have retained here, too, the character of a dynamic polarization. . . . When Einstein, then, made the statement of his turn from "I" and "We" to "It," he testified only to the vividly existential sense with which he experienced a certain polarity between work and intimacy. This, as a certain imbalance, has been described in a number of studies as typical to scientists.[34]

I shall return to this polarity in section 4 below, after a look at Einstein's thought processes in section 3. Hopefully it will then be clear why periods of isolation were so essential to his development.

2. The Technological Milieu in Which Einstein Grew Up

2a. The Einstein Family Business

Einstein's sister, Maja, gives a good account of how Einstein's father got into electrotechnology:

A younger brother of Hermann Einstein named Jakob, who later was to exert a certain intellectual influence on the growing Albert, had finished his engineering studies and decided to set up a firm [in Munich] devoted to water conduction and electrical equipment. Since his own resources would not have sufficed for this, he induced his brother Hermann to become a partner in the business, both personally as the business manager as well as through a considerable financial investment. . . . At a time when the whole world began to introduce electric lighting, the modestly begun enterprise seemed to be developing well. But Jakob Einstein's plans aimed much higher. His many-sided and fertile ideas led him among other things to the construction of a dynamo machine of his own invention that he wanted to manufacture on a large scale. It required a larger manufacturing plant and significant capital to set his plans in motion. The entire family, especially Hermann's father-in-law Julius Koch, made this possible by financial participation, and the enterprise came into being.[35]

We are fortunate to have the reminiscences of one of the firm's employees, Aloys Höchtl.[36] When Höchtl went to work for the J. Einstein & Co. Electrotechnical Factory in 1886, it had six lathes, eighteen workbenches, and two forges.

As is possible only in a smaller workshop, every day I was confronted by different types of work. . . . [E]verything was done in our own workshop. . . . I soon mastered the construction of dynamos of all types with the greatest assurance. I was entrusted with the construction of control panels, arc lamps and measuring instruments, so that I knew about every aspect of electric lighting plants (there were still no power transmission installations).[37]

Höchtl discusses technical progress made by the firm in dynamo construction and electric lighting and their mass production, which for a time kept J. Einstein & Co. competitive with the largest German firms such as Siemens & Halske, Allgemeine Elektrizitäts-Gesellschaft (AEG), and S. Schuckert.[38] From other sources we know that the firm ultimately came to employ between 50 and 200 people—still a small firm but far from minuscule. In October 1886, the firm was entrusted with the installation of electric lighting for the Munich *Oktoberfest*, a major event in the city's social life, and the lighting of various local firms, such as the

Pschorr Brewery (then as now, beer played a large role in Munich).

By the end of the 1880s, orders began to flow in, a number from abroad. Jakob Einstein, sometimes alone and sometimes with his foreman, Sebastian Kornprobst, began to take out patents on dynamos, lighting installations, and electrical measuring instruments. Patents were taken out not only in Germany, but in Italy and even one in the United States! Orders for electrification of the street lighting in smaller cities began to come in, notably from Varese in northern Italy and Schwabing, now a suburb of Munich. Then came the firm's big opportunity:

> In early 1893, the city of Munich announced the construction of electric street lighting amounting to about 360 arc lamps with a power generating plant of 300 horsepower = 200 kW. Einstein & Co presented a project that had some prospects for implementation. But finally the project of the firm S. Schuckert of Nuremberg was chosen, that was based on its experience in the previous execution of large installations.
>
> For the owners of the firm J. Einstein & Co, Mssrs. Jacob and Hermann Einstein, this was a heavy blow, which they experienced as particularly mortifying, because the Munich firm did not get the chance, but rather the execution was entrusted to a Nuremberg firm [also in Bavaria]. The firm S. Schuckert was rich in experience with just such large installations and in addition was much more heavily endowed with capital.
>
> A plan to erect a branch factory in Italy, already considered earlier as a consequence of active business connections there, was renewed. As a consequence of the general decline in business conditions in 1893 [a year of economic crisis], which necessitated the dismissal of some workers as well as a reduction of working hours, this plan was strengthened to a deci-

sion to move the whole business to Milan. This decision did not remain unknown to us, and caused a general uneasiness among the personnel.[39]

Looking back in 1899 at the development of electrotechnology in Munich, two of its leading figures, Oskar von Müller and Prof. E. Voit, summed up the story:

> It would have surely developed more easily if a significant electrotechnical factory had been located here. . . . The initially favorably developing firm of J. Einstein & Co. was driven away by the more powerful external firms that soon set up subsidiaries in Munich.[40]

Caught up in the movement at that time by electrical firms not only to build but operate power stations by raising the necessary capital themselves, the Einstein brothers soon decided to erect a large factory and operate a power station in Pavia, a city south of Milan, using the waters of the Naviglio to power the generator. However, as we learn from the Pavia newspapers of the time, they soon became involved in a financial imbroglio. A local cooperative was set up to manage the city power supply, but the Einstein firm secretly tried to buy up the rights to the waters of the Naviglio, hoping to present the cooperative with a fait accompli. When the cooperative found this out, they canceled their contract with the Einstein firm and gave the electricity rights to another company.

As a result, the Einstein brothers were forced to liquidate their firm in the summer of 1896, the start of a period of financial hardships and worries for Hermann Einstein that only ended with his death. Maja Einstein reports:

> Not only were the assets of Albert Einstein's mother lost at this time, but significant contributions from relatives as well.

The family had hardly anything left. [Brother Jakob went to work as an engineer for another firm.] In contrast, Albert Einstein's father could not bring himself to take the same step and relinquish his professional independence. In particular, he did not want to bring suffering on his wife, who would have had great difficulty accommodating herself to a lower standing in the social scale. Against the perceptive advice of his still quite young son, he founded a third electrical firm in Milan.[41]

Again, relatives provided the needed funding; this time it was Rudolf Einstein, the father of Einstein's cousin Elsa, who later became his second wife. This small firm had its ups and downs, and never really brought financial security to the family. The Einsteins' precarious situation seems to have affected Hermann's previously robust health, and in 1902 he died of heart disease, still in debt to his brother-in-law Rudolf when his business was liquidated.[42]

2b. Einstein's Reaction to It

What was the effect of so many years of living in close contact with the electrotechnical business on the young Einstein? He was only a year old when the family moved to Munich so that his father could join Jakob's business, fifteen when the family moved to Italy for business reasons, and twenty-three when his father died a business failure. I suggest that its effect was to give rise to another polarity in his personality: On the one hand, he was *attracted* to the *technical* aspects of the business, to which he even contributed from time to time. On the other, he was *repelled* by its *commercial* aspects, which led to one disappointment after the other of his family's hopes for financial success, or finally just financial independence. Let us look at these two aspects, turning first to technology.

Some evidence has survived of Einstein's youthful interest in toys that involved hands-on construction, and in learning how technical devices work. His sister reports that, before the age of ten, his games—which were "very indicative of his natural aptitudes"—included "work with the fret-saw and the erection of complicated structures with the well-known 'Anker' building sets, but best of all the construction of many-storied houses of cards."[43] A fellow student at the Luitpold Gymnasium recalled Einstein explaining to him the principles of the telephone. Uncle Jakob had dabbled (unsuccessfully) in marketing an early model of the telephone, and, at a time when they were still a rarity, there was one in the Einstein home, so one may surmise that is how he learned about its workings.

Somewhat later, Uncle Jakob remarked to Sebastian Kornprobst, "Do you know, it is really fabulous with my nephew. Whereas my assistant engineer and I broke our heads a whole day long, the young lad worked out the whole story in a mere quarter of an hour. Something more will come of him!"[44] From his first relativity paper, we learn that Einstein was familiar with the debate over the problem of unipolar induction,[45] a debate closely connected with the engineering problems of electric dynamos.[46] Since design and construction of electric dynamos was a major activity of the Einstein firm and Uncle Jakob actually held patents on one dynamo design, it is again a fair surmise that young Albert first heard of the problem in this engineering context.

In his authorized biography, Einstein's stepson-in-law Rudolf Kayser wrote, "As a result of his father's calling and his own mathematical ability, the position of technician and engineer was the first to be thought of [for Einstein]."[47] But Kayser immediately adds:

> The choice of profession, however, had other implications: it made necessary a relationship with society and with a mechanical life constantly controlled by end-in-view and utilitarian purposes. Nothing seemed more frightful to young Albert Einstein. Moreover he was not ambitious: he wanted neither fame nor success. These mundane ideas were repugnant to him.
>
> [As for the Einstein firm, d]isastrous commercial failure began all over again. The boy was once more surrounded by discussions and cares which seemed to belong to another world. He still showed a strong dislike for action and haste which valued material profit. The more pressing the need for a vocational activity became, the greater was his antipathy against it. He wanted only to see, to comprehend and to experience. The world, however, did not seem to approve of this very desire. The laws of society were antagonistic to the character of the thoughtful young dreamer.[48]

In September 1896, after the Italian business had collapsed, Einstein wrote (in his French Examination Essay for the *Matura*) about his hopes of studying mathematics and physics and then becoming a "teacher [*professeur*] of these branches of natural science, choosing the theoretical part of these sciences." His reasons were "above all, the personal [*individuel*] disposition for abstract thought and mathematics, the lack of imagination [*fantaisie*] and of practical talent. . . . Then, it is also a certain independence of the scientific profession that pleases me very much."[49]

The family business worries continued to plague him. He wrote his sister, Maja, in 1898 about their father's decision to start his own new business rather than to seek employment in another firm:

> If things had gone according to my ideas, papa would already have looked for a position two years ago [1896], so that he and we would have been spared the worst. . . . What depresses me most of course is the misfortune of my poor parents who have not had a happy moment for so many years. What further hurts me is that as an adult I have to look on without being able to do anything. After all, I am nothing but a burden to my family.[50]

Indeed, his family was unable to support him financially while he attended the Poly, and he lived on a monthly stipend from the wealthy maternal side of the family, the Kochs. Einstein still felt obliged, however, to help out in his father's business:

> I thought I could make better use of my vacation time by studying a few things of importance and by learning about my father's business in Italy. It is quite possible, after all, that he could suddenly take ill, or become otherwise indisposed, and he has no one else to turn to.[51]

For a while, things seemed to be going better: "My father has become quite a different fellow since his money troubles are over. You can see that all the dark clouds have really vanished from the fact that he will make a trip to Venice with me after both of us have visited his power plants."[52] But by 1901, the troubles were back: "The poor things [his parents] have constant troubles and cares on account of the damned money. My dear uncle Rudolf (the rich) has been nagging them terribly [Rudolf was Hermann Einstein's chief creditor]."[53] His father died a year later.

In later years Einstein summarized his attitude toward

technology and business: "I was also originally supposed to be a technical worker. But the thought of having to expend my inventive power on things, which would only make workaday life more complicated, with the goal of dreary oppression by capital, was unbearable to me."[54] Notice his use of the phrase "inventive power." As we shall see in the next section, rather than use the term "discovery," Einstein preferred to borrow the term "invention" from its original technological milieu and apply it to any creative intellectual process.

Einstein found solace in his job at the Swiss Patent Office for many reasons, but one of them seems to be rooted in the positive side of his ambivalent attitude toward technology: his post provided him with an opportunity to think about interesting technical issues without taking any responsibility for their commercial application. How comfortable he felt there can be seen from a letter he wrote to his wife about a year after he started work:

> I am getting along with Haller [the Director of the Patent Office] better than ever. He is quite friendly and recently, when a Patent agent complained about my decision, and into the bargain on the basis of a decision of the German Patent Office, he agreed I was right on every point. You'll see, I'll get ahead now so that we won't have to starve.[55]

A fascination with inventions continued throughout Einstein's life. Starting with his work in 1907 with the Habicht brothers on the "Maschinchen,"[56] he took out a number of patents, alone or jointly with others; and he was consulted as a patent expert in a number of important cases. He was never a worshiper of technology, though, always remaining sensitive to its social implications:

Concern for technology is suited in great measure to counteract a degeneration of science . . . [into sterile formalism]. . . .
[O]ne must make technology into a true cultural factor by
bringing its rich intellectual and aesthetic content closer to
public awareness. What comes to the mind of a sensitive person upon hearing the word technology? Greed, exploitation,
social division among the people, class hatred, mindless rest-
less striving. . . . [I]s it any wonder that the educated friend of
humanity hates technology as a misguided child of our age,
that threatens to destroy the finer pleasures of life? In order
that this robust child of society be directed towards salvation,
we must not let it grow up wild. One must seek to understand
it, in order to gain influence over it. It has power over forces
that could ennoble life.[57]

After this excursion into his technological interests, it will
perhaps not surprise the reader to learn in the next section
that Einstein's thought processes centered around visual and
muscular elements rather than verbal activities.

3. What, Precisely, Is "Thinking"?

Einstein asks this question in his *Autobiographical Notes*,
and I advise the reader to consult his answer in full, as well
as the essay on this topic by Gerald Holton.[58] Of course, the
Autobiographical Notes were written when Einstein was
sixty-seven, and he reminds us that "today's person of sixty-
seven is by no means the same as was the one of fifty, of
thirty, or of twenty. Every reminiscence is colored by one's
present state, hence by a deceptive point of view."[59] But he
adds; "Nevertheless much can be gathered out of one's own
experience that is not open to another consciousness."[60] I
shall try to be cautious in singling out a few points of rele-

vance from this and other later writings of Einstein to add to the contemporary evidence from the period up to 1905, to argue for two theses:

1. For Einstein, the process of thinking was a solitary activity, primarily non-verbal in nature. At a secondary stage, it was necessary for him to transform the results of this primary process into forms communicable to others.

2. Although thinking is an solitary activity, the need to put ideas into a communicable form led Einstein to search throughout his early years (indeed throughout his life) for people to act as "sounding boards" for his ideas. "Sounding board" is a phrase that Einstein once used to characterize the role of his friend Michele Besso. As I have explained elsewhere, "it means that Besso was capable of understanding things that Einstein explained to him, and of asking intelligent questions that could help Einstein develop his own ideas . . . but that Besso was not capable of any creative effort of his own."[61]

Now to the evidence for the first thesis. After Einstein and his assistant Ernst Straus had jointly solved a certain problem, Einstein expressed what Straus called his "very strongly held conviction":

This has never happened to me—we thought together. Two people being able to think! I never thought that this was possible. I was always convinced that thinking was a singular occupation.[62]

This comment was made quite late in Einstein's life (Straus was his assistant from 1944 to 1948), and since he first published two jointly authored papers as early as 1908,[63] he clearly had something different in mind than simple collaboration in the writing of a paper. So I take it that he meant just what he said: that the kind of thinking that led to new ideas was a solitary activity for him. This is confirmed, curiously enough, by a passage in Banesh Hoffmann's account of his collaboration with Einstein and Leopold Infeld:

> When we got into a situation where we did not know what to do next . . . and this happened at least three times, . . . Einstein would say, "I will a little think" . . . [a]nd then he would twirl his hair like this and he would walk up and down or stand still and his face had no sign of strain at all. He seemed as if he were in another part of the universe, only his body being with us, and Infeld and I kept absolutely silent. I simply do not know how long this went on. There was Einstein thinking like this and after a while he would suddenly relax, come back to earth, look at us, smile, and say, Yes we should do such and such. Of course, it worked, and that is how we got out of those very deep difficulties.[64]

What were the non-verbal and verbal steps in this thinking process? We don't have much information about this from his early years, but we do have one important clue. His speech developed rather later than is the norm: "It is true that my parents were concerned, because I began to speak comparatively late, and that they consulted their physician for that reason."[65] But by his third year he had begun to speak. He told Ernst Straus that

> when he was between two and three years old, he formed the ambition to talk in whole sentences. If somebody asked him a question and he had to answer, he would form a sentence in his mind and then try it out on himself, thinking that he was whispering it to himself. But, as you know, a child is not very good at whispering, so he said it softly. Then, if it sounded all right, he would say it again to the person who had questioned him. Therefore, it sounded, at least to his nursemaid [note that the family could afford a nursemaid!], as if he said every-

thing twice, once softly and once loudly, and she called him "der Depperte," which is Bavarian for "the dopey one." The nickname stuck, and that, at least in Einstein's mind, was the cause of all the stories about his slow development.[66]

According to his sister, Maja, "this curious habit persisted until his seventh year."[67]

But the role of auditory sensations in his verbal processing persisted. He told Robert Shankland, "'I am the acoustic type; I learn by ear and give by word. When I read I hear the words. Writing is difficult and I communicate this way very badly.' . . . He told me [Shankland] that he even hated to write his *Autobiographical Notes* in German."[68]

I shall now turn to Einstein's comments on the nature of his thought process, written in 1945. If read in light of his late development of speech and his emphasis on the sound of words, they become more understandable:

It is indubitable to me, that our thinking occurs for the greatest part without the utilization of signs (words), and in addition also largely without consciousness. . . . It is in itself not necessary that a concept be attached to a sign (word) that is perceptible to and reproducible by the senses; however, if it is, then thinking thereby becomes communicable.

Words or language, as they are written or spoken, do not seem to play any role in my mechanism of thought. The psychical entities which seem to serve as elements in thought are certain signs and more or less clear images which can be "voluntarily" reproduced and combined. . . . [T]he above mentioned elements in my case are of visual and some of muscular type. Conventional words or other signs [in his case, probably mathematical] have to be sought for laboriously only at a secondary stage. . . . In a stage when words intervene at all, they are in my case purely auditory.[69]

Indeed, Einstein worried about the tendency of education to convert all conceptual thinking into verbal forms:

> Education continually threatens a danger peculiar to it, the severing of the link to sensory experience. Every educational process creates a world of concepts. At their inception, the latter are closely tied to the realities, for the lucid grasping of which they were created. But there is a tendency to universalization attached to linguistically fixated concepts, which on the one hand extends their range of applicability, on the other hand weakens their connection with a sensory experience. . . . Who will deny that the *Gymnasia*, in which the center of interest is directed towards the linguistic, are exposed to this danger in an especially great measure?[70]

Many of the crucial thought experiments Einstein later reports confirm the existence of this first stage of the thinking process, utilizing visual and muscular imagery. I shall mention just four:

1. chasing after a light ray at the speed of light;
2. moving a magnet toward a conductor versus moving the conductor toward the magnet;
3. falling freely under the influence of gravity and feeling nothing untoward;
4. visualizing a gravitational wave through its effect on an elastic body.

The first two thought experiments are connected with the development of the special theory of relativity, the last two with the development of the general theory. Let us listen to his testimony on each of these:

> [1.] After ten years of reflection such a principle resulted from a paradox upon which I had already hit at the age of sixteen: If I pursue a beam of light with the velocity c (velocity

of light in vacuum), I should observe such a beam of light as an electromagnetic field at rest though spatially oscillating. There seems to be no such thing, however, neither on the basis of experience nor according to Maxwell's equations. From the very beginning it appeared to me intuitively clear that, judged from the standpoint of such an observer, everything would have to happen according to the same laws as for an observer who, relative to the earth, was at rest. For how should the first observer know, or be able to establish, that he is in a state of fast uniform motion?

One sees that in this paradox the germ of the special relativity theory is already contained. Today everyone knows, of course, that all attempts to clarify this paradox satisfactorily were condemned to failure as long as the axiom of the absolute character of time, or of simultaneity, was rooted unrecognized in the unconscious.[71]

[2.] It is well known that Maxwell's electrodynamics . . . when applied to moving bodies, leads to asymmetries that do not seem to be inherent in the phenomena. Take, for example, the electrodynamic interaction between a magnet and a conductor. The observable phenomenon here depends only on the relative motion of conductor and magnet, whereas the customary view draws a sharp distinction between the two cases, in which either the one or the other of the two bodies is in motion.[72]

[3.] Then there came to me the most fortunate thought of my life in the following form: The gravitational field only has . . . a relative existence, like that of an electric field generated by electromagnetic induction. Because for an observer freely falling from the roof of a house, during his fall there exists— at least in his immediate neighborhood—no gravitational field. Indeed if the observer lets go of any sort of body, then relative to him it remains in a state of rest or uniform motion independently of its particular chemical and physical nature. The observer is thus justified in interpreting his state as being at rest.[73]

A visitor from Göttingen, Rudolf Humm, reports that in 1917 Einstein told him:

[4.] He works more imaginatively and does not seem to trust the work we are doing in Göttingen; he has never thought on such formalistic lines. His powers of imagination are closely related to reality. He told me that he visualizes the gravitational waves with the help of an elastic body, and at the same time he made a movement with his fingers as though he were pressing an india rubber ball.[74]

I suggest that it was during a secondary stage in his thought process, when the need for communication arose, that exposition of his ideas to others, and the discussion of such ideas, became—if not a necessity for Einstein, then at least a highly desirable activity. Of course, I do not mean to suggest that the second stage followed the first in rigid sequence; rather, I suggest that he moved back and forth between the two stages in the course of the development of his ideas. In short, there was a dialectical tension between the "invention" (for this term, see the discussion below) of his ideas in creative solitude and the communicative elaboration of these ideas with the help of his sounding boards.

I believe that these two stages together constitute what Einstein meant by the words "grübeln," "Grübelei" [ruminating, rumination (or brooding)] that he used often to characterize the process by which his ideas developed. When once asked about the nature of his genius, he demurred at the term, but insisted on his curiosity and ability to persevere in such ruminations over the wonders that he turned up (I shall come back to "wonder" in a moment). He told Ernst Straus, "Two things are needed for our kind of work: indefatigable perseverance and readiness to take something, into which one has put much time and effort, and throw it away."[75]

He felt that the chief task of the scientist is to find the most important question and then to pursue it without deviating from the main problem. "You must never let yourself be seduced by any problem, no matter how difficult it is."[76] In reminiscences of his years at the Poly, Einstein states:

[P]hysics . . . was divided into separate fields, each of which was capable of devouring a short lifetime of work without having satisfied the hunger for deeper knowledge. The mass of insufficiently connected experimental data was overwhelming here also. In this field, however, I soon learned to scent out that which might lead to fundamentals, and to turn aside from everything else, from the multitude of things that clutter up the mind and divert it from the essentials.[77]

This is presumably what Einstein meant when he said, "Scientific greatness is essentially a question of character. The main thing is: Don't make any rotten compromises."[78]

One of these "most important questions" over which Einstein brooded is the relation between light and the laws of mechanics. He ruminated on this for about ten years before arriving at what we now call the special theory of relativity. He also brooded over the nature of light for fifty years but, up to the end of his life, and in spite of his extraordinary contributions to the quantum theory of light, was never satisfied with the answers that he or any of his contemporaries were able to give.[79]

Now I shall return to the question of the significance of a "wonder" to Einstein. As Erik Erikson emphasized, Einstein had the ability to keep the child alive in the man, so let us turn back to the child. The earliest wonder Einstein recalled was that "as a child of four or five years . . . my father showed me a compass."[80]

What made something a wonder for Einstein? The answer is not unconnected with the prior point, as seen here in the continuation of a previously cited passage from the *Autobiographical Notes*:

> It is indubitable to me, that our thinking occurs for the greatest part without the utilization of signs (words), and in addition also largely without consciousness. For how, otherwise, should it often happen that quite spontaneously we "are surprised at (wonder at)" some occurrence? This "surprise at (wonder at)" appears to arise when an occurrence comes into conflict with a conceptual world that is sufficiently fixated within us.[81]

I interpolate here a sentence that actually comes later:

> What a person sees before him from infancy onwards he does not react to in this way; one does not wonder at the falling of objects, at the wind and the rain, nor at the moon nor the difference between the animate and the inanimate.[82]

So, for Einstein, it is not some inherent "wonderfulness" of the phenomenon that makes for "wonder," but an apparent conflict between this phenomenon and our established conceptual framework. A tree, for example, no matter how beautiful, is not a wonder in Einstein's sense—but a *talking* tree would be. Now we can understand why the compass was a "wonder" to young Albert.

That this needle behaved in such a definite way did not at all fit into the kind of occurrences that could find a place in my unconscious conceptual world (actions connected with touching or contact). I can still remember—or believe myself to remember [the man of sixty-seven remembering the five-year-old]—that this occurrence made a deep and lasting impression on me. There must be something behind things, which is deeply hidden.[83]

Resuming the earlier text:

When such a conflict is experienced harshly and intensively, then it reacts back upon our conceptual world. The development of this conceptual world is in a certain sense a continual flight from "surprise" ["Wunder"—Einstein here plays on the double sense of the German word].[84]

The 1905 relativity paper (see p. 124 of this volume) presents such a conflict between the postulate

that not only the phenomena of mechanics but also those of electrodynamics have no properties that correspond to the concept of absolute rest. Rather, the same laws of electrodynamics and optics will be valid for all coordinate systems in which the equations of mechanics hold . . . [which postulate] will hereafter be called "the principle of relativity"

and the

only seemingly incompatible [postulate] . . . that light always propagates in empty space with a definite velocity V that is independent of the state of motion of the emitting body.

The first (relativity) postulate is suggested by the phenomenon of electromagnetic induction discussed in the paragraph on the magnet and the conductor, "together with the unsuccessful attempts to detect a motion of the earth relative to the 'light medium'"; while the second (light) postulate is also based on a wealth of experimental results: the phenomenon of aberration, Arago's experiment, Fizeau's experiment, etc.[85] The conflict between these two sets of phenomena is what causes the "wonder" in Einstein's sense,[86] which "reacts back upon our conceptual world," which must be modified in such a way that the conflict is shown to be "only apparent."

Speaking of the "basic concept of relativity," Einstein said, as reported by Moszkowski:

It is not correct that this basic principle came to me as a primary concept. If it had arisen in this way, then there might be justification for calling it a "discovery." But just the suddenness that you presuppose must be denied. Rather I was led to it *stepwise* by *individual* regularities (occurrences in accord with natural laws) taken from experience.[87]

That is why Einstein preferred the word "invention" to "discovery" to characterize his work: "Invention occurs here as a constructive activity."[88]

Turning finally to the question of his "sounding boards," Einstein evidently felt the need throughout his life to expound his ideas to sympathetic listeners—even listeners who had no training as physicists. Ernst Straus reports that

he explained all his new ideas to his sister, who only had a relationship to physics through him. However, she must have been an excellent listener, because he often said: 'Yes, my sister also agrees. . . .' It is, I believe, this ability to reduce profound concepts to their intuitive content that was in large part the cause of his popular fame.[89]

The first people to play such a role in his life were his uncle Jakob, who introduced him to algebra and posed him technical problems (see the previous section), and Max Talmey, a young medical student, eleven years Albert's senior, who introduced him to popular-scientific literature and started to discuss scientific and philosophical issues with the ten-year-old boy.[90] During his years at the Poly in Zurich, Marcel Grossmann and Michele Besso played important roles in the process of Einstein's clarification of his ideas through dialogue with others.[91] And, as we shall see in the next section, during these years, he attempted to combine both creative work and intimacy in his relationship with Mileva Marić, an attempt that ultimately failed.

4. EINSTEIN AND MILEVA MARIĆ

At the end of section 1 I cited Erik Erikson on the role of "a certain polarity between work and intimacy" in Einstein's life. Through his surviving correspondence, we are able to locate evidence of this polarity—or tension—surprisingly early in his life. As is the case with most young people, Einstein's search for intimacy was closely tied to the search for love. The young Einstein was attractive to women and knew how to use his boyish charm, even manifesting a flirtatious streak.[92]

The first serious love relationship that he had was with Marie Winteler, one of the three daughters of Jost and Pauline Winteler, with whom he boarded while attending the Aargau Kantonsschule (see section 1).[93] They became surrogate parents with whom he established close and lasting relationships, referring to them as "Mama" and "Papa" (note that both his real and surrogate mothers were named Pauline). He fell in love with Marie at the age of seventeen. She was a couple of years older—a recurrent pattern in his love life—and studying at the local teachers training second-ary school, after which she took a position at a nearby primary school.

The depth of their relationship may be gauged from this extract from one of his letters to her, written from his parents' home in Italy during a school break:

> I have now had to see in fullest measure, my little angel, what homesickness and yearning mean. Yet how much more happiness love brings than yearning brings pain. Now I first realize how indispensable to my happiness my dear little sun has become. . . . You are more to my soul than the entire world used to be.[94]

She responded with equal passion, and both sets of parents approved of the relationship.[95] After graduating from the Kantonsschule, he moved to Zurich to attend the Polytechnic and, despite Marie's ardent letters,[96] the move led to an emotional as well as a physical separation from Marie. One year after the letter cited above, Einstein wrote to "Mama" Winteler, declining an invitation to spend the Easter holidays with the Winteler family:

> It would be more than shameful if I were to buy a couple of days of pleasure at the price of the pain, much too much of which I have already caused the dear child through my fault. It fills me with a sort of strange satisfaction to have to experience now a part of the pain that my thoughtlessness and ignorance have caused such a delicate nature as that of the dear girl. Strenuous intellectual work and the contemplation of God-given Nature are the angels that will guide me—reconciled, fortified, and yet inexorably strong—through all the turmoil of this life. . . . One thus creates for himself a small world—however pitifully insignificant it may be compared to the eternally changing greatness of true existence—and yet

feels wonder himself at how great and important it is, just like for example the mole in the hole that he has excavated for himself.[97]

It is interesting to compare this statement with something he wrote over twenty years later:

[O]ne of the strongest motives that leads to art and science, is escape from everyday life with its painful crudity and bleak aridity, from the fetters of ever-changing personal desires. . . . Each human being tries to make for himself in the fashion that suits him best a simplified and intelligible picture of the world and thus to overcome the world of experience, by trying to some extent to substitute for it this picture of his. . . . He displaces the center of gravity of his emotional life to this picture and its shaping in order to seek the repose and solidity that he cannot find in the all-too-narrow sphere of the maelstrom of personal experience.[98]

Although expressed here in more sophisticated words, one sees that his attitude had not changed basically since 1897.

While Einstein's relationship with Marie Winteler is in no way unusual as adolescent love affairs go, several motifs that will recur over and over in his emotional life—notably with both his wives—already appear in this episode: his longing for a deeply intimate relationship with a woman; his tendency sooner (as in this case) or later (as in his two marriages) to withdraw emotionally from the relationship after attaining his goal; and his flight from the "merely personal" into an ever-more-successful attempt to center his emotional life around the creation of an "extra-personal" world, in which he can take refuge.

Of course, neither does this make Einstein unique: there must be millions of people today with similar emotional histories. What constitutes his greatness is what he was able to accomplish within this "small world" or "simplified picture of the world" of his own, and how much of his vision he was able to impose on "the eternally changing greatness of true existence" or "the world of experience."

His relationship with Marie Winteler never encompassed complete intimacy nor was there ever a question of joint work in physics. It does not seem to have been consummated sexually (she later wrote: "We loved each other deeply, but it was a completely ideal love."); and although Winteler was a primary school teacher, she seems to have had neither interest in nor inclination to the study of physics. But Einstein soon attempted to unite the two poles of work and intimacy through the bonds of joint study of physics and sexual love that for a time united him with Mileva Marić. She was three years older than he and a fellow student in the small group of students entering the mathematics-physics teachers section VIA of the Poly.

Elsewhere I have told the story of how this attempt to maintain intimacy and a working relationship in physics failed and ultimately came to grief.[99] Here I shall confine myself to the years up to 1905. It is particularly important to address their relationship during these years, since a recent television program broadcast by PBS, now available on DVD and with a Web site devoted to it,[100] makes the claim that "Maric [*sic*], a brilliant mathematician, collaborated with [Einstein] on three famous works: Brownian Motion, Special Relativity Theory and Photoelectric Effect."[101] If this statement were correct, I would be honor bound to include Marić's name on the title page of this book and to explain her role in this collaboration. In the appendix to this section, I hope to show that there is no valid evidence supporting such claims.

As noted above, Einstein's parents had approved of Marie Winteler, but became more and more hostile toward Marić. It seems clear that he used his relationship with her to help him break free from his parents—in particular from the dominant role that his mother tried to play in his life. At any rate, initially their union was a deeply felt one. Einstein's early letters to Marić include many examples of how he extended the boundaries of his ego to include her, a not-uncommon phenomenon between intimate lovers. Here are two such examples:

> I'm so lucky to have found you, a creature who is my equal [*ebenbürtig*] and who is just as strong and independent as I am myself! Except with you, I am alone with everyone.

> You must always be my witch and my street-urchin. . . . Everyone but you seems alien to me, as if they were separated from me by an invisible wall.[102]

He even became dependent on joint study with her: "Even studying is only half the enjoyment, without you."[103] Although his letters to Marić make it clear that he was using their relationship to help him break free of his family,[104] he evidently felt unable to marry until he got his father's deathbed blessing in 1902. And this was in spite of the fact that, earlier in that year, Marić gave birth to their daughter, referred to as "Lieserl" in the letters. Lieserl's fate is still unknown, but her existence has itself given birth to the wildest speculations. One Einstein expert even hazarded the opinion that Marić and Einstein practiced birth control up to the point at which Marić stopped doing so in order to have a child out of wedlock! One can hardly believe he has actually read the letters between them: the constant notes of real despair in Marić's letters during her pregnancy accompanied by the equally constant reassurances by Einstein of his love, and reiterated assurances in his letters that, in the long run, all will turn out well, that they will be able to marry after he lands a job; and that in him, she has more than any of her girlfriends, who are already starting to get married.[105]

Initially, Einstein also tried to draw Marić into his creative life. His letters to her are full of accounts of his ideas about physics, including new theoretical ideas and proposals for new experiments. I have discussed these letters in detail elsewhere,[106] and so will merely state here that, even in the case in which we have Marić's direct response to Einstein's letter detailing his rather striking ideas about the electrodynamics of moving bodies, no response to his ideas on this subject, or any other topic in physics that he raised, is found in her letters.

Yet she certainly listened attentively to what he said. Einstein's friend at the Poly, Margarete von Uexküll (see section 1 above), who also knew Marić well (for some time they lived in the same boarding house), made a most interesting comment:

> [Einstein] had the ability to explain difficult problems with clarity, she recalled, and would pour out his ideas as they walked home from the laboratories. "I believe," said von Uexküll, "that Mileva was the first person who believed in his theories. When I once remarked that I found Einstein's theories quite fantastic, she answered confidently: 'But he can prove his theory.' Quietly, I thought to myself, she must really be in love !"[107]

I believe that, originally, Einstein wanted to share an intellectual collaboration with Marić, but that she simply could not rise to his expectations. She certainly played the role of a "sounding board" for his ideas, a role that Michele Besso also played in this period, and that many others were to play in the course of Einstein's life (see section 3 above). Marić undoubtedly played the role of sympathetic—even emphathetic—listener, vital to the secondary process by which Einstein transformed his non-verbal thoughts into communicable words (see section 3 above), and passionate advocate and defender of his ideas when he had no direct access to the scientific community. She also served from time to time as his amanuensis and probably helped him look up data and check calculations. But there is no evidence at all that she contributed ideas of her own to his creative process, let alone that they took part in a joint one (recall Einstein's amazement when something like this happened late in his life—see the beginning of section 3 above).

Whether it might have been possible for a different, more talented, more assertive woman—a Marie Curie or a Tatiana Ehrenfest—to play a more collaborative role in the young Einstein's creative activities; or whether his talents required intellectual solitude, it is impossible to say for sure, although if I had to put my money on one of the two alternatives I would put it on the latter. What is certain to me—after a careful examination of the supposed evidence to the contrary (see the appendix)—is that there is no evidence that Mileva Marić ever did play such a role.

As I have shown in detail elsewhere, the places in his letters to Marić where Einstein refers to "our work" are quite general statements; when it comes to specific assertions about the work, he invariably uses the first person singular ("I," "my," etc.) in describing it. I think two factors must be considered when evaluating these general statements: The inclusion of Marić within the boundaries of his own ego during the early, most intense phase of their love affair sometimes led him to ignore the distinction between "mine" and "ours."[108] And when, during her pregnancy out of wedlock and living apart from Einstein, Marić gave voice to her despair, as often happened, he answered reassuringly, picturing their future work together.[109] Here is one example:

> When you are my dear little wife, we'll diligently work on science together, so we don't become old philistines, right?[110]

But this collaboration never came to pass. After their marriage, even the role she had played during their student years began to diminish. Their son Hans Albert Einstein, when asked "And your mother, how did she treat his rising fame?" replied,

> She was proud of him, but that is as far as it went. It was very hard to understand, because she had originally studied with him and had been a scientist herself. But, somehow or other, with the marriage she gave up practically all of her ambitions in that direction.[111]

In "Einstein and Marić," I have discussed possible reasons for the loss of her earlier ambitions, and Einstein was certainly not guiltless in this respect. Even if her talents did not match his own, he could have involved her more in his intellectual work. But the fact is that, after settling in Bern, Einstein developed another "charmed circle" of sounding boards. It consisted of two of his close friends, Maurice Solovine and Konrad Habicht, who together with Albert constituted the entire membership of their self-created *Akademie Olympia*, which they set up to mock the real Academies of Sciences, from which they were excluded. This "charmed circle" soon included Michele Besso once again, his friend from Zurich school days, whom Einstein helped to get a job at the Patent Office, and a few other friends.[112]

By 1905, while not entirely excluded from the "charmed circle" (she is reported to have sat silently but attentively through many sessions of the *Akademie Olympia*), Marić was definitely at the fringes, and the Einstein-Marić relationship had begun to take on the "philistine" attributes that Einstein had earlier been so anxious to avoid. After they settled down together in Bern, he wrote his friend Besso:

> Well, I am now an honorably married man, and lead a fine comfortable life with my wife. She takes care of everything exceptionally well, cooks well, and is always cheerful.[113]

Marić wrote to her friend Helene Savić:

> I am, if possible, even more attached to my dear treasure than I already was in the Zurich days. He is my only companion and society and I am happiest when he is beside me.[114]

Her inquiry in this letter about possible teaching jobs for Albert and herself in Belgrade, where Savić had settled with her husband, is Marić's last known reference to the possibility of a career in physics.

By 1905 their relationship had settled into a sort of equilibrium, which lasted until it was exposed to the kind of contact with the larger world that started when Einstein left the Patent Office in 1909 to assume his first full-time academic position at the University of Zurich. His emotional detachment became more and more pronounced as his scientific reputation grew, and Marić was well aware of what was happening:

> He is now counted among the leading German-speaking physicists and is being frightfully courted. I am very happy

about his success, which he has really earned: I only hope and wish that fame does not exert a detrimental influence on his human side.[115]

APPENDIX

Now I must reluctantly turn to the claim, cited at the beginning of the last section, that "Maric [sic], a brilliant mathematician, collaborated with [Einstein] on three famous works: Brownian Motion, Special Relativity Theory and Photoelectric Effect." Believe me, it is not with any pleasure that I turn to this task. I am fully aware of the truth in the observation of a nineteenth-century Danish diplomat "Denials never have the charm or the impact of false reports." But this book would not appear in Einstein's name alone if there were any credible evidence for such claims.

The claim that "Maric [was] a brilliant mathematician" must be set against the fact that she took the final examinations at the Poly twice, and each time failed because of her low grades in mathematics. But this would be beside the point if we had any evidence that she "collaborated with [Einstein] on three famous works: Brownian Motion, Special Relativity Theory and Photoelectric Effect." The only evidence offered for the claim on the PBS Web site is the following statement:

> But there is at least one printed report in which Joffe ["Abram Joffe (Ioffe), a respected member of the Soviet Academy of Sciences"] declared that he personally saw the names of two authors on the 1905 papers: Einstein and Marity (a Hungarianized form of Maric).[116]

What evidence is given for this claim? On the same page of the Web site is an illustration of a part of a text in Russian, with the caption: "Old Russian journal citing Einstein-Marity (Maric) as co-authors of the 1905 papers." No sources are cited for this claim beyond the illustration.

In fact, the illustration is *not* from an article by Joffe, *not* from "an old Russian journal," and does *not* cite Einstein and Marić as "co-authors of the 1905 papers." The "printed report" by Joffe is a 1955 article in the Soviet journal *Uspekhi fizicheskikh nauk*,[117] which also does *not* cite Einstein and Marić as "co-authors of the 1905 papers."

The illustration is actually from page 57 of a popular-science book by Daniil Semenovich Danin, *Neizbezhnost strannogo mira*, published in 1962 by the Molodaia Gvardii [Young Guard] publishers in Moscow. The text of the relevant passage reads, in English translation:

> The unsuccessful teacher, who, in search of a reasonable income, had become a third class engineering expert in the Swiss Patent Office, this yet completely unknown theoretician in 1905 published three articles in the same volume of the famous "Annalen der Physik" signed "Einstein-Marity" (or Marić—which was his first wife's family name).

This English translation is taken from a book by Christopher Jon Bjerknes.[118] The Russian text is reproduced on page 196 of the book, and comparison with the illustration on the Web site, which is the same as that used on the PBS television program, establishes that this is indeed the "old Russian journal" cited.

Now, the articles were presumably not signed "Einstein-Marity" *and* Einstein-Marić—so which was it? Danin obviously has no clue how they were signed. He is merely amplifying a story picked up somewhere else—indeed from Joffe, as we shall see in a moment. As far as I know, Danin's text was first cited in print in connection with a discussion of Einstein and Marić by Margarete Maurer.[119] Neither Bjerknes nor Maurer can be regarded as biased in favor of Albert Einstein (I would say quite the contrary), but she states, "The page copied from Danin's work still does not indeed represent a historical 'proof,'" and proceeds to suggest that it probably originated from Joffe's reminiscences, which were not available to her at the time.

Indeed, if one looks at the passage from Joffe, it is clear that it is the source of Danin's assertions, so let us turn to this passage. In English translation, it reads:

> For physics and especially for the physics of my generation—that of Einstein's contemporaries, Einstein's entrance into the arena of science is unforgettable. In 1905, three articles appeared in the "Annalen der Physik", which began three very important branches of 20th century physics. Those were the theory of Brownian motion, the photon theory of light, and the theory of relativity. The author of these articles—an unknown person at that time, was a bureaucrat at the Patent Office in Bern, Einstein-Marity (Marity—the maiden name of his wife, which by Swiss custom is added to the husband's family name).

Again, the translation (slightly modified in the descriptions of the three papers) is from the book by Bjerknes, pages 195–196; the Russian original is given on page 196. We see that all Danin adds to the statements by Joffe is the well-known fact that Einstein was a "third class engineering expert" at the Patent Office; and the addition after Einstein-Marity, "(or Marić—which was his first wife's family name)," a circumstance to which we shall return below.

Why did the producers of the PBS show and the authors of the Web site choose to show an excerpt from Danin and not from Joffe? One can only speculate; but it is noteworthy that, although he says nothing about two authors, Danin does use the phrase "signed" while Joffe does not. The text cited above is *all there is* in the original article by Joffe that bears on the question of authorship of the three 1905 articles. To summarize, he states that their author was someone working at the Swiss Patent Office, whose name was "Einstein-Marity" (note again that the phrase "or Marić" does *not* occur in Joffe).

All further claims about Joffe on the PBS program and Web site are actually based on the assertions of Desanka Trbuhović-Gjurić in her biography of Marić:

> The distinguished Russian physicist Abraham F. Joffe (1880–1960) pointed out in his "In Remembrance of Albert Einstein" that Einstein's three epochal articles in Volume 17 of "Annalen der Physik" of 1905 were originally signed "Einstein-Marić."[120]

Note that this is *not* what Joffe says, which is: "The author of these articles . . . was . . . Einstein-Marity." He makes no claim about the signature, and certainly not about having *seen* the original signature. But let us go on with the quotation from Trbuhović-Gjurić:

> Joffe had seen the originals as Assistent to Röntgen, who belonged to the Board of the "Annalen," which had examined submitted contributions for editorial purposes. Röntgen showed his *summa cum laude* student this work, and Joffe thereby came face to face with the manuscripts, which are no longer available today.[121]

Trbuhović-Gjurić offers no documentation of, or other justification for, her two claims that (1) Röntgen had the original manuscripts, and that (2) he then showed them to Joffe; nor, when she was later interviewed by Robert Schulmann, one of the editors of the Einstein Papers, was she able to offer any further evidence beyond the reference to a microfilm of an article, which her son subsequently stated was the "article" by Danin (see the discussion in Maurer, cited above). Note that, if the first claim by Trbuhović-Gjurić fails, the second fails as an immediate consequence. So let us start by looking at the second claim, for the moment "bracketing" the truth of the first.

If the second claim were correct, it is hard to see why Joffe never mentioned this most interesting and unusual fact during the fifty-five years between 1905 and his death in 1960. Why did he not mention it in the 1955 article? And why not in his autobiographical memoir,[122] which has a chapter on Einstein? But he never claimed in print that he saw the original manuscript—nor did anyone else until *after* Joffe's death. If the memory stayed so vivid in his mind that he remembered the name "Einstein-Marity" correctly fifty years after having last seen it, why did he never mention the fact, even when he published his two reminiscences of Einstein? The simplest explanation of why he never claimed to have seen the manuscripts is that he never did.

Of course the very possibility of the second claim depends entirely on the validity of the first. If Röntgen had examined the three papers in manuscript form, it is curious that he waited until September 1906 to write to Einstein asking for reprints of his papers on electrodynamics to add to his (Röntgen's) collection of papers on this topic. He adds that he has been concerned with Brownian motion for some time and is thus familiar with Einstein's work on this topic, and does not ask for a reprint of this work. This contrast suggests that Röntgen was not familiar with Einstein's work on electrodynamics in late 1906, casting doubt on the first part of Trbuhović-Gjurić's claim. And why should Paul Drude, the editor of the *Annalen* in 1905 and the author of two books and numerous articles on electromagnetic theory and optics, have needed to call upon Röntgen, an experimentalist in Munich, to vet Einstein's purely theoretical manuscripts? That Drude was familiar with, and positively evaluated, Einstein's electrodynamics paper is evident from Drude's citations of the relativity paper twice in print before he died in 1906.

Drude's advisor on theoretical physics papers for the *Annalen* was Max Planck, whom Einstein's sister credits with sending Einstein the first written reaction from a physicist to his 1905 relativity paper.[123] Here is what the foremost account of the development of theoretical physics in Germany has to say about the editorial practices of the *Annalen der Physik* during the years around 1905:

> At the same time [1894] he [Planck] acquired an official responsibility for theoretical physics for all of Germany, replacing Helmholtz as the designated advisor on matters of theoretical physics for the *Annalen der Physik*. . . . When in 1900 Drude became editor of the *Annalen der Physik*, Planck continued on as advisor. Their working relationship was good, even if Planck was not always kept as informed as he wished.

As the advisor on theoretical physics for the *Annalen der Physik*, in 1905 Planck was already familiar with Einstein's work. For five years, Einstein had regularly submitted papers to this journal, the most important of which treated thermodynamics and statistical physics, subjects of particular interest to Planck at the time. Einstein extended these studies to a related interest of Planck's, black body radiation, in 1905.

Einstein's relativity theory of the same year set Planck to work; it was the subject, Max Born observed, that "caught Planck's imagination more than anything else."[124]

When confronted with these circumstances, defenders of the claims put forward by Trbuhović-Gjurić usually counter with the question: How else could Joffe have known that "Marity" was the form of her name that Marić sometimes used? For example, Evan Walker Harris (one of those interviewed on the PBS program) states that "Ioffe would only have known had he seen the original signed by her, since this usage of 'Mariti' apparently is not to be found in any of the Einstein biographies."[125] This last statement is incorrect. The second edition of Carl Seelig's well-known biography of Einstein, published the year before Joffe's article, gives her name as "Mileva Maric [*sic*] oder Marity"[126]

But whether or not "this usage" of Marity is to be found in any other of the Einstein biographies, there could well be other published sources, in which Joffe could have found this fact. "Marity" is, after all, the form in which her name appears on their marriage certificate, and this fact might have been picked up by some other writer about Einstein. A careful search would have to be conducted of the literature about Einstein in several languages, including Russian, before a valid judgment could be passed on the question of where Joffe could have found the information. I myself would hazard the guess that Joffe saw this form of her name in some document published right after Einstein's death. If he had only seen it some fifty years earlier, it is hard to explain how he was able to reproduce the exact form of the name in 1955.

But even aside from a source in print, there is always the

possibility that he heard it from someone. One intriguing possible explanation of how Joffe might have heard it is from Mrs. Einstein herself! In Joffe's book of reminiscences,[127] there is a chapter on Einstein. It does *not* include the story about having seen "Einstein-Marity" on the 1905 papers; but it *does* include an account of a meeting with Einstein's wife in 1905.

> I wanted very much to talk to Einstein about all these questions and, together with my friend Wagner, visited him in Zurich. But we did not find him at home and so did not succeed in talking to him. However his wife told us that, as he himself expressed it, he is only a Patent Office clerk, and cannot think seriously about science, much less about experiments.[128]

This is a curious story in many ways: when he worked at the Patent Office, Einstein and Marić were in Bern of course and not Zurich. And if Marić really reported his own words to them as cited here, it would appear that she was being ironic. Of course, it is possible that Joffe remembered incorrectly many details of a visit that had taken place fifty-odd years earlier. But if there is any grain of truth to the story (and if we are going to deny any grain of truth to Joffe's stories, we might as well throw out the first one about "Einstein-Marity" along with this one), then he may have picked up "Marity" from Mrs. Einstein herself during their conversation—or even from the name on the apartment doorbell—and later added some garbled information about Swiss customs for the names of married couples.

Perhaps even more likely is the possibility that Joffe learned about the name Marity from Paul Ehrenfest. He was friendly with Ehrenfest for decades, and the published correspondence between them covers the years from 1907 until the latter's untimely death.[129] Ehrenfest had been well acquainted with both Einstein and Marić since about 1911 or 1912, and could easily have been the source of some information about Marić's name that Joffe remembered years later in garbled form.

But let us suppose—contrary to all these arguments—that Trbuhović-Gjurić's two claims *were* valid. How do we pass from these claims—that the papers had *one* signature (Einstein-Marity)—to the claim that this *one* signature represents *two* authors? The three papers in question contain many authorial comments in the first person *singular*. One example from each paper follows (italics added):

> In this paper *I* wish to present the train of thought and cite the facts that led *me* onto this path . . . (this volume, p. 178).

> It is possible that the motions to be discussed here are identical with so-called Brownian molecular motion; however, the data available to *me* on the latter are so imprecise that *I* could not form a judgment on the question (this volume, p. 85).

> In conclusion, let *me* note that *my* friend and colleague M. Besso steadfastly stood by *me* in *my* work on the problem discussed here, and that *I* am indebted to him for several valuable suggestions (this volume, p. 159).

Of course, this does not settle the question of who did the work. But it does show that the articles were written with one authorial voice, and so does seem to settle the question of whether the work was submitted with two named authors—unless we are to believe that the editors of the *Annalen* not only removed one of the named authors, but carefully changed all uses of the first person plural to first person singular!

We have seen that, in order to give credence to Trbuhović-Gjurić's claims, we are forced to pile one improbability upon another: the improbability of Röntgen having had the manuscript, the improbability that Joffe saw it, the improbability that his assertion that the papers were written by one person should be interpreted as meaning they were written by two people. The simplest and most natural course is to reject *all* of these implausible claims.

John Stachel
January 2005

EDITORIAL NOTES

[1] Jürgen Renn and Robert Schulmann, eds., *Albert Einstein/Mileva Marić: The Love Letters* (Princeton, N.J.: Princeton University Press, 1992), cited hereafter as *Love Letters*, p. 67. Given in the original German as "Es lebe die Unverfrorenheit! Sie ist mein Schutzengel in dieser Welt" in *The Collected Papers of Albert Einstein* (Princeton University Press, 1987–), cited hereafter as *Collected Papers*, vol. 1, p. 323.

[2] *Collected Papers*, vol. 5, p. 3. The original German reads, "Ich suche die Einsamkeit, um mich dann still über sie zu beklagen." All translations, unless otherwise indicated, are mine.

[3] See the 3 January 2000 issue of *Time*.

[4] In fact, not one of the three covers of *Time* that have featured Einstein since 1929 has depicted the young Einstein.

[5] For further discussions of a number of topics raised both here and in the original introduction, see John Stachel, *Einstein from 'B' to 'Z'* (Boston: Birkhäuser, 2002).

[6] Gerald Holton seems to have been the first person to have noted "a remarkable set of puzzling polarities . . . in Einstein's style and life's work." See Holton, "On Trying to Understand Scientific Genius," in *Thematic Origins of Scientific Thought*, rev. ed. (Cambridge, Mass.: Harvard University Press, 1988), pp. 371–398; citation from p. 374. For a discussion of the concept of polar opposition, see John Stachel, "The Concept of Polar Opposition in Marx's *Capital*," to appear in Stachel, *Going Critical*, vol. 1., *The Challenge of Practice* (Boston: Kluwer Academic, 2005).

[7] Erik Erikson, "Psychoanalytic Reflections on Einstein's Centenary," in G. J. Holton and Y. Elkana, eds., *Albert Einstein, Historical and Cultural*

Perspectives: The Centennial Symposium in Jerusalem (Princeton, N.J.: Princeton University Press, 1982), p. 152.

[8] Ibid., p. 153. Einstein's only sister, Maja, who was two years younger, reports that she was the first victim of his rages, including attacks with a skittle-ball and a children's pickaxe (*Collected Papers*, vol. 1, p. lvii).

[9] All schools in Bavaria were denominational, and there was no Jewish school in Munich at the time.

[10] Albert Einstein, letter of 3 April 1920, *Collected Papers*, vol. 9, p. 492; translation cited from John Stachel, "Einstein's Jewish Identity," in *Einstein from 'B' to 'Z'*, p. 59. This article gives more details about anti-Semitism in his classroom.

[11] Letter of 25 January 1918, *Collected Papers*, vol. 8B, p. 614.

[12] Ernst Straus reports that Einstein told him, "When I was at the *Gymnasium* in Munich, my class teacher came to me and said: 'I would be very happy if you were to leave us.' I responded: 'But I haven't been guilty of any wrong-doing.'— 'Yes, but your very presence in the class is enough to destroy all respect.'" See Ernst Straus, "Assistent bei Albert Einstein," in Carl Seelig, ed., *Helle Zeit—Dunkle Zeit/ In Memoriam Albert Einstein* (Zurich: Europa Verlag, 1956), p. 73.

[13] *Collected Papers*, vol. 1, p. lxiii.

[14] To avoid confusion, note that the Aargau Kantonsschule is located in the town of Aarau, Switzerland.

[15] *Collected Papers*, vol. 2, p. 11.

[16] Carl Seelig, *Albert Einstein/Eine dokumentarische Biographie*, 2d ed. (Zurich: Europa, 1954), p. 22.

[17] G. J. Whitrow, ed., *Einstein, The Man and His Achievement* (London: British Broadcasting Corporation, 1967; New York: Dover Publications, 1973), p. 4. Citations are to the Dover edition.

[18] Roger Highfield and Paul Carter, *The Private Lives of Albert Einstein* (New York: St. Martin's Press, 1994), pp. 39–40. Einstein's assessment of Pernet appears to have been correct: He was already a sick man, and died a few years later.

[19] *Collected Papers*, vol. 1, p. 47.

[20] See below and "Einstein as a Student of Physics, and His Notes on H. F. Weber's Course," in *Collected Papers*, vol. 1, pp. 60–62.

[21] See Carl Seelig, *Albert Einstein/A Documentary Biography* (London: Staples Press, 1956), p. 30. Seelig gives further evidence of Weber's hostility toward Einstein.

[22] Philipp Frank, *Einstein, His Life and Times*, rev. ed. (New York: Knopf, 1953), pp. 20–21. The book was prepared with Einstein's coopera-

tion. The translation has been revised using the German original, *Einstein, Sein Leben und seine Zeit*, first published in 1949 and later in 1979, with a foreword by Einstein.

[23] Marić to Helene Kaufler, 4 June–23 July 1900, *Collected Papers*, vol. 1, pp. 244–245. One of Marić's girlfriends, Milana Bota, wrote her mother on 7 June 1900, "I see little of Mitza [Marić's nickname] on account of her German, whom I hate" (ibid.). I will continue this discussion of the Einstein-Marić relationship in section 4.

[24] Einstein, "Autobiographische Skizze," in Seelig, *Helle Zeit—Dunkle Zeit*, p. 10.

[25] See "H. F. Weber's Lectures on Physics," in *Collected Papers*, vol. 1, pp. 63–210.

[26] Letter of 16 February 1898, *Collected Papers*, vol. 1, p. 212.

[27] She had been away from the Poly for a semester, and consequently took the exams a year after Einstein.

[28] Einstein to Mileva Marić, 28? September 1899, *Love Letters*, pp. 15–16.

[29] Albert Einstein, *Autobiographical Notes*, Paul Arthur Schilpp, ed. and trans. (LaSalle, Ill.: Open Court, 1979), p. 17. This is a corrected edition of an earlier-published version of a work written in 1947.

[30] Ibid., p. 15.

[31] Hermann Einstein to Wilhelm Ostwald, 13 April 1901, *Collected Papers*, vol. 1, p. 289.

[32] Einstein to Mileva Marić, 7? July 1901, *Collected Papers*, vol. 1, p. 308, and *Love Letters*, pp. 56–57.

[33] Einstein himself later stated: "I am a horse for single harness, not cut out for tandem or team work. I have never belonged wholeheartedly to country or state, to my circle of friends, or even to my own family. These ties have always been accompanied by a vague aloofness, and the wish to withdraw into myself increases with the years" ("Albert Einstein," in *Living Philosophies* [New York: Simon and Schuster, 1931], p. 4. Another translation of this chapter appears as "The World As I See It," in *Ideas and Opinions* [New York: Crown, 1954], pp. 8–11.).

[34] "Psychoanalytic Reflections," pp. 157–158. Frank's descriptions of Einstein are taken from the biography *Einstein, His Life and Times* (see note 22).

[35] Maja Winteler-Einstein, "Albert Einstein—Beitrag für sein Lebensild (Excerpt)," in *Collected Papers*, vol. 1, pp. l–li.

[36] Aloys Höchtl, "*Lebenserinnerungen* von Aloys Höchtl, geschrieben München 1934" (unpublished manuscript). Excerpts are included in

INTRODUCTION, CENTENARY EDITION

Nicolaus Hittler, *Die Elektrotechnische Firma J. Einstein u. Cie in München—1876–1894* (n.p., n.d.). Dokument Nr. 11512 aus den Wissens-archiven von Global Research and Information Network, found at www.grin.de. Unless otherwise indicated, Hittler's thesis is the source for all information cited on the Einstein family business while it was located in Munich.

[37] Ibid., p. xii.

[38] These were the three other firms that bid for the Munich electric street-lighting contract, discussed below.

[39] Höchtl, *"Lebenserinnerungen,"* in Hittler, *Die Elektrotechnische*, p. xvi.

[40] Oskar von Müller and Dr. E. Voit, "Elektrotechnik in München," in *Die Entwicklung Münchens unter dem Einflüsse der Naturwissenschaften während der letzten Dezennien—Festschrift der 71. Versammlung deutscher Naturforscher und Aerzte gewidmet von der Stadt München* (n.p., 1899), p. 132.

[41] Winteler-Einstein, "Albert Einstein—Beitrag für sein Lebensild," p. liv.

[42] See *Collected Papers*, vol. 1, p. lv, note 31.

[43] Winteler-Einstein, "Albert Einstein—Beitrag für sein Lebensild," p. lix.

[44] Otto Neustätter to Einstein, 12 March 1928, cited in *Collected Papers*, vol. 1, p. lxiv.

[45] He shows that the problem disappears in the relativistic framework. See Albert Einstein, "Zur Elektrodynamik bewegter Körper," *Annalen der Physik* 17 (1905): 891–921, reproduced in *Collected Papers*, vol. 2, pp. 276–306. Unipolar induction is mentioned on p. 295.

[46] See Arthur I. Miller, "Unipolar Induction: A Case Study of the Interaction between Science and Technology," *Annals of Science* 38 (1981): 155–189, reprinted in Arthur I. Miller, *Frontiers of Physics, 1900–1911: Selected Essays* (Boston: Birkhäuser, 1986), pp.153–189.

[47] Anton Reiser [Rudolf Kayser], *Albert Einstein: A Biographical Portrait* (New York: Albert and Charles Boni, 1930), p. 42.

[48] Ibid., pp. 42, 43.

[49] *Collected Papers*, vol. 1, p. 28.

[50] *Collected Papers*, vol. 1, p. 211.

[51] Einstein to Mileva Marić, 14? August 1900, *Collected Papers*, vol. 1, pp. 254–255, and *Love Letters*, pp. 26–27.

[52] Einstein to Mileva Marić, 20 August 1900, *Collected Papers*, vol. 1, pp. 255–257, and *Love Letters*, p. 28.

[53] Einstein to Mileva Marić, 23 March 1901, *Collected Papers*, vol. 1, pp. 279–281, and *Love Letters*, p. 38.

[54] Einstein to Heinrich Zangger, before 11 August 1918, *Collected Papers*, vol. 8B, p. 850.

[55] Einstein to Mileva Einstein-Marić, 19? September 1903, *Collected Papers*, vol. 5, p. 22.

[56] See "Einstein's 'Maschinchen' for the Measurement of Small Quantities of Electricity," in *Collected Papers*, vol. 5, pp. 51–55.

[57] Einstein, "Die Freie Vereinigung für technische Volksbildung. Eine Zuschrift des Professors Dr. Einstein an die Vereinigung. Wien, 23. Juli [1920]," *Neue Freie Presse*, 24 July 1920, *Morgen-Ausgabe*, p. 8, reproduced in *Collected Papers*, vol. 7, p. 336. An immediately preceding quotation from this article is given in the next section.

[58] Gerald Holton, "'What, precisely, is 'thinking?' . . . Einstein's answer" in A. P. French, ed., *Einstein: A Centenary Volume* (Cambridge: Harvard University Press, 1979), pp. 153–164.

[59] *Autobiographical Notes*, p. 3.

[60] Ibid.

[61] John Stachel, "The Young Einstein: Poetry and Truth," in Stachel, *Einstein from 'B' to 'Z'*, p. 36.

[62] Ernst Straus, "Reminiscences," in Holton and Elkana, *Albert Einstein, Historical and Cultural Perspectives*, p. 420.

[63] The two papers were written with Jacob Laub. See *Collected Papers*, vol. 2, doc. 51–53.

[64] Banesh Hoffman, "Working With Einstein," in Harry Woolf, ed., *Some Strangeness in the Proportion: A Centennial Symposium to Celebrate the Achievements of Albert Einstein* (Reading, Mass.: Addison-Wesley, 1980), pp. 477–478.

[65] Einstein to Sybille Blinoff, 21 May 1954, cited in *Collected Papers*, vol. 1, p. lvi, note 35.

[66] Straus, "Reminiscences," p. 419.

[67] Einstein-Winteler, "Albert Einstein—Beitrag für sein Lebensild," p. lvii.

[68] Robert S. Shankland, "Conversations with Albert Einstein," *American Journal of Physics* 31 (1963): 50. I am indebted to Alberto Martinez for this reference.

[69] These comments, in response to an enquiry by the noted mathematician Jacques Hadamard, are printed in Hadamard's book *An Essay on the Psychology of Invention in the Mathematical Field* (Princeton, N.J.: Princeton University Press, 1945), and reprinted as "A Mathematician's

INTRODUCTION, CENTENARY EDITION

Mind" in *Ideas and Opinions*, pp. 25–26. The bracketed comments are mine.

[70] Einstein, "Die Freie Vereinigung für technische Volksbildung," p. 338. The continuation of this quotation is given in the previous section (see note 57).

[71] *Autobiographical Notes*, pp. 49, 51.

[72] "On the Electrodynamics of Moving Bodies," in this volume, p. 123.

[73] "Grundgedanken und Methoden der Relativitätstheorie, in ihrer Entwicklung dargestellt," in *Collected Papers*, vol. 7, p. 265.

[74] Diary of Rudolf Jakob Humm, cited in Seelig, *Albert Einstein/A Documentary Biography*, p. 155.

[75] Straus, "Assistent bei Albert Einstein," p. 70.

[76] Ernst Straus, "Memoir," in A. P. French, ed., *Einstein: A Centenary Volume* (Cambridge, Mass.: Harvard University Press, 1979), p. 31.

[77] Einstein, *Autobiographical Notes*, p. 15.

[78] Straus, "Assistent bei Albert Einstein," p. 72.

[79] See John Stachel, "Einstein and the Quantum: Fifty Years of Struggle," in Robert Colodny, ed., *From Quarks to Quasars: Philosophical Problems of Modern Physics* (Pittsburgh: University of Pittsburgh Press, 1986), pp. 349–385; reprinted in Stachel, *Einstein From 'B' to 'Z'*, pp. 367–402.

[80] *Autobiographical Notes*, p. 8. The story is *first* told in Alexander Moszkowski, *Einstein, Einblicke in seine Gedankenwelt* (Hamburg: Hoffmann and Campe, 1921), p. 219, based on conversations with Einstein.

[81] *Autobiographical Notes*, pp. 6, 8. The translation is my own, and keeps both senses of the German word *Wundern*.

[82] Ibid., p. 8.

[83] Ibid.

[84] Ibid.

[85] For a discussion of these phenomena, see Michel Janssen and John Stachel, "The Optics and Electrodynamics of Moving Bodies," preprint 265 of the Max-Planck-Institut für Wissenschaftsgeschichte, to appear in Stachel, *Going Critical*, vol. 1 (see note 6); and John Stachel, "Fresnel's Dragging Coefficient as a Challenge to 19th Century Optics of Moving Bodies," preprint 281 of the Max-Planck-Institut für Wissenschaftsgeschichte, to appear in the *Proceedings of the Sixth International Conference on the History of General Relativity, Amsterdam 2002*.

[86] In "Grundgedanken" (see note 73) Einstein says, "The phenomenon of electromagnetic induction drove me to postulate the principle of special relativity" (p. 265), but adds in a footnote: "The difficulty to be overcome

INTRODUCTION, CENTENARY EDITION

lay in the constancy of the speed of light in the vacuum, which at first I thought had to be abandoned. Only after years of fumbling about did I realize that the difficulty rested upon the arbitrariness of the basic kinematical concepts" (p. 280, n. 34).

[87] Moszkowski, *Einstein*, p. 103.

[88] Ibid., p. 101. The German word for "invention" is "Erfindung." On the previous page, Einstein decisively rejects the term "discovery" ("Entdeckung"): "Discovery is really not a creative act."

[89] Straus, "Assistent bei Albert Einstein," p. 71.

[90] See Max Talmey, "Formative Period of the Inventor of Relativity Theory," in *The Relativity Theory Simplified and the Formative Period of Its Inventor* (New York: Falcon Press, 1932), Part III, pp. 159–179. Talmey comments: "For five years I had the good fortune of associating very often with the young mathematician and philosopher. In all these years I never saw him reading any light literature. Nor did I ever see him in the company of any schoolmates or other boys of his age. He usually held himself aloof, absorbed in books on mathematics, physics, and philosophy" (ibid., pp. 164–165).

[91] See Einstein, "Autobiographische Skizze," pp. 9–17 (see note 24 above), and the references to Grossman and Besso in *Collected Papers*, vols. 1 and 5.

[92] Two early examples: A poem inscribed in the album of the young daughter of the innkeepers of the hotel where Einstein was staying (*Collected Papers*, vol. 1, p. 220); and the inscription on his photograph, given to an Italian friend: "To Madame, but not to Miss, Marangoni" (see Illustration 19 in *Collected Papers*, vol. 1).

[93] For information about Marie Winteler, see the biographical sketch "Müller-Winteler, Marie," in *Collected Papers*, vol. 1, p. 385; for information on her parents, see "Winteler, Jost, Winteler-Eckart, Pauline," in ibid., p. 388.

[94] Einstein to Marie Winteler, 21 April 1896, *Collected Papers*, vol. 1, p. 21.

[95] This may be one of the reasons why he soon ended it. As we shall see, he used his parents' opposition to his relationship with Mileva Marić to help liberate himself from their influence.

[96] See Marie Winteler to Einstein, 4–25 and 30 November 1896, *Collected Papers*, vol. 1, pp. 50–53.

[97] Einstein to Pauline Winteler, May 1897, *Collected Papers*, vol. 1, pp. 55–56.

[98] Albert Einstein, "Motive des Forschens" [Motives for Research] (speech given in 1918 at a meeting in honor of Max Planck's fiftieth birth-

day), in *Collected Papers*, vol. 7, pp. 55–56. The English translation, "Principles of Research" in *Ideas and Opinions*, pp. 224–227, citations from p. 225, has been modified.

[99] John Stachel, "Albert Einstein and Mileva Marić: A Collaboration That Failed to Develop," cited hereafter as "Einstein and Marić," in Helena M. Pycior, Nancy G. Slack, and Pnina G. Abir-Am, eds., *Creative Couples in the Sciences* (New Brunswick, N.J.: Rutgers University Press, 1996), pp. 207–219, 330–335; reprinted in Stachel, *Einstein from 'B' to 'Z'*, pp. 39–55.

[100] *Einstein's Wife*, PBS DVD Video B8958, described on the cover as "The story of Einstein's secret marriage and scientific collaboration." http://www.pbs.org/opb/einsteinswife/

[101] Quotation from the cover of the DVD. The Web site makes a more modest claim: "There are several credible scientists who believe Mileva may have collaborated on some at least of the 1905 papers."

[102] Einstein to Mileva Marić, *Love Letters*, 3 October 1900, p. 36, and 28 December 1901, p. 73, respectively.

[103] Einstein to Mileva Marić, 17? February 1902, *Love Letters*, p. 76.

[104] See, for example, *Love Letters*, pp. 19–20.

[105] See, for example, his letters of 28 November, 12 December, and 19 December 1901, *Love Letters*, pp. 68–71.

[106] See "Einstein and Marić," and also "The Young Einstein: Poetry and Truth" and "Einstein and Ether Drift Experiments," both in Stachel, *Einstein from 'B' to 'Z'*.

[107] Highfield and Carter, *Private Lives of Einstein*, p. 40.

[108] Here one thinks of the beautiful passage from Shakespeare's *The Merchant of Venice* in which Portia reveals her love to Bassano: "One half is yours, the other half yours—mine own I would say, but if mine then yours, and so all yours" (act 3, scene 1).

[109] Again, see "The Young Einstein: Poetry and Truth" and "Einstein and Ether Drift Experiments" in Stachel, *Einstein from 'B' to 'Z'* for the details.

[110] Einstein to Mileva Marić, 28 December 1901, *Love Letters*, pp. 72–73.

[111] Whitrow, *Einstein, The Man and His Achievement*, p. 19.

[112] For the Bern years, see Max Flückiger, *Albert Einstein in Bern* (Bern: Paul Haupt, 1974).

[113] Einstein to Michele Besso, 22 January 1903, "Einstein and Marić," p. 41.

[114] Marić to Helene Savić, 20 March 1903, "Einstein and Marić," pp. 41–42.

[115] Marić to Helene Savić, 3 September 1909, "Einstein and Marić," p. 42.

[116] http://www.pbs.org/opb/einsteinswife/science/mquest.htm. Unless otherwise noted, the source for all the following citations is the PBS Web site http://www.pbs.org/opb/einsteinswife/.

[117] A. F. Joffe, "Pamyati Alberta Eynshtyna," *Uspekhi fizicheskikh nauk* 57 (1955). I cite this article from the reprint in *Eynshtyn i sovremmenaya fizika. Sbornik pamyati Eynshtyna* (Moscow: GTTI, 1956), pp. 20–26; the reference to "Eynshtyn-Mariti" is on p. 21. I thank Dr. Gennady Gorelik for his help in finding this reference.

[118] *Albert Einstein: The Incorrigible Plagiarist* (Downers Grove, Ill.: XTX Inc., 2002), p. 197.

[119] In an article entitled "Weil nicht sein kann, was nicht sein darf . . . 'DIE ELTERN' ODER 'DER VATER' DER RELATIVITÄTSTHEO-RIE," which originally appeared in Birgit Kanngiesser et al., eds., *Dokumentation des 18. Bundesweiten Kongresses von Frauen in Naturwissenschaft und Technik vom 28.–31. Mai 1992 in Bremen* (Bremen: n.p., n.d.), pp. 276–295, since reprinted in various versions and available at http://www.rli.at/Seiten/kooperat/maric1.htm for the first part (to access the second and third parts, and the bibliography, substitute the numbers 2, 3, and 4, respectively, for the number 1 in the Web address). This is the version that I am citing.

[120] Bjerknes, *Albert Einstein: The Incorrigible Plagiarist*, p. 197. The German text from which this was translated appears on p. 198 and is taken from Desanka Trbuhović-Gjurić, *Im Schatten Albert Einsteins/Das tragische Leben der Mileva Einstein-Marić* (Bern: Paul Haupt, 1983), which is a German translation of the Serbian original.

[121] Ibid.

[122] *Vstrechi s fizikami, moi vospominaniia o zarubezhnykh fizikah* [Meetings with Physicists, My Reminiscences of Physics Abroad] (Moscow: Gusudarstvennoye Izdatelstvo Fiziko-Matematitsheskoi Literatury, 1962).

[123] See *Collected Papers*, vol. 2, p. xxx.

[124] Christa Jungnickel and Russell McCormmach, *Intellectual Mastery of Nature*, vol. 2, *The Now Mighty Theoretical Physics, 1870–1925* (Chicago: University of Chicago Press, 1986), pp. 254–255, 309, and 248.

[125] "Mileva Marić's Relativistic Role," letter in *Physics Today* (February 1991): 122.

[126] Seelig, *Albert Einstein/Eine dokumentarische Biographie*, p. 29. The English edition, with the corresponding passage, did not appear until 1956; see Seelig, *Albert Einstein/A Documentary Biography*, p. 24.

[127] See note 122. I have consulted the German edition, *Begegnungen mit Physikern* (Leipzig: B. G. Teubner, 1967).

[128] Ibid., pp. 88–89.

[129] See *Ehrenfest-Ioffe Nauchnaya perepiska, 1907–1933* (Leningrad: Nauka, 1973). Dr. Gennady Gorelik kindly informed me that there is no mention of Mileva Marić in this correspondence.

PUBLISHER'S PREFACE

In 1905, Einstein produced five of his most significant contributions to modern science, all of which first appeared in the prestigious German journal *Annalen der Physik* in that year. More recently, they have reappeared in the original German, with editorial annotations and prefatory essays, in volume 2 of the *Collected Papers of Albert Einstein*, an ongoing series of volumes being prepared by the Einstein Papers Project at Boston University under the sponsorship of Princeton University Press and the Hebrew University of Jerusalem.

Einstein's Miraculous Year draws heavily from this volume (*The Swiss Years: Writings, 1900–1909*), which remains the definitive and authoritative text of all of Einstein's writings of those years; we encourage scholars to consult it when seeking original texts and detailed discussions and annotations of Einstein's work. For the present volume, we have compiled Einstein's five major papers of 1905 and included, in abridged form, the historical essays and notes that deal with his contributions to relativity theory, quantum mechanics, and statistical mechanics and adapted them for presentation in this special edition. We are therefore indebted to the editors of volume 2 for their scholarly contributions: John Stachel, David C. Cassidy, A. J. Kox, Jürgen Renn, and Robert Schulmann.

The English translations that appear here are new. The intention has been to render Einstein's scientific writings accurately into modern English, but to retain the engaging and clear prose style of the originals. We are deeply grateful to Trevor Lipscombe, Alice Calaprice, Sam Elworthy, and John Stachel for preparing them. We also thank Dr. Attila Piroth, translator of the Hungarian edition, for his helpful list of errata and misprints in the original edition.

Einstein's Miraculous Year

INTRODUCTION

I

To anyone familiar with the history of modern science, the phrase "miraculous year" in the title immediately calls to mind its Latin counterpart "annus mirabilis," long used to describe the year 1666, during which Isaac Newton laid the foundations for much of the physics and mathematics that revolutionized seventeenth-century science. It seems entirely fitting to apply the same phrase to the year 1905, during which Albert Einstein not only brought to fruition parts of that Newtonian legacy, but laid the foundations for the break with it that has revolutionized twentieth-century science.

But the phrase was coined without reference to Newton. In a long poem entitled *Annus Mirabilis: The Year of Wonders, 1666,* John Dryden, the famed Restoration poet, celebrated the victory of the English fleet over the Dutch as well as the city of London's survival of the Great Fire. The term was then used to celebrate Newton's scientific activities during the same year—a year in which he laid the foundations of his version of the calculus, his theory of colors, and his theory of gravitation.[1] Here is Newton's own (much later) summary of his accomplishments during this period:

> In the beginning of the year 1665 I found the Method of approximating series & the Rule for reducing any dignity

[power] of any Binomial into such a series [i.e., the binomial theorem]. The same year in May I found the method of Tangents . . . , & in November had the direct method of fluxions [i.e., the differential calculus] & the next year in January had the Theory of colours & in May following I had entrance into [th]e inverse method of fluxions [i.e., the integral calculus]. And the same year I began to think of gravity extending to [th]e orb of the Moon & (having found out how to estimate the force with w[hi]ch [a] globe revolving within a sphere presses the surface of a sphere [i.e., the centrifugal force]): from Kepler's rule of the periodical times of the Planets being in sesquialterate proportion of their distances from the centers of their Orbs [i.e., Kepler's third law], I deduced that the forces w[hi]ch keep the Planets in their Orbs must [be] reciprocally as the squares of their distances from the centers about w[hi]ch they revolve: & thereby compared the force requisite to keep the Moon in her Orb with the force of gravity at the surface of the earth, & found them answer pretty nearly. All this was in the two plague years of 1665 & 1666. For in those days I was in the prime of my age for invention & minded Mathematicks & Philosophy more then [sic] at any time since.[2]

More recently, the term *annus mirabilis* has been applied to the work of Albert Einstein during 1905 in an effort to establish a parallel between a crucial year in the life of the founding father of classical physics and of his twentieth-century successor.[3] What did Einstein accomplish during his miraculous year? We are fortunate in having his own contemporary summaries of his 1905 papers. Of the first four he wrote to a close friend:

4

I promise you four papers . . . , the first of which I could send you soon, since I will soon receive the free reprints. The paper deals with radiation and the energetic properties of light and is very revolutionary, as you will see. . . . The second paper is a determination of the true sizes of atoms from the diffusion and viscosity of dilute solutions of neutral substances. The third proves that, on the assumption of the molecular [kinetic] theory of heat, bodies of the order of magnitude of 1/1000 mm, suspended in liquids, must already perform an observable random movement that is produced by thermal motion; in fact, physiologists have observed motions of suspended small, inanimate, bodies, which they call "Brownian molecular motion." The fourth paper is only a rough draft at this point, and is an electrodynamics of moving bodies, which employs a modification of the theory of space and time; the purely kinematical part of this paper will surely interest you.[4]

Einstein characterized the fifth paper in these words:

One more consequence of the paper on electrodynamics has also occurred to me. The principle of relativity, in conjunction with Maxwell's equations, requires that mass be a direct measure of the energy contained in a body; light carries mass with it. A noticeable decrease of mass should occur in the case of radium. The argument is amusing and seductive; but for all I know, the Lord might be laughing over it and leading me around by the nose.[5]

The parallels are clear: each man was in his mid-twenties; each had given little previous sign of the incipient flowering of his genius; and, during a brief time span, each struck out on new paths that would ultimately revolutionize the science of his times. If Newton was only twenty-four in 1666 while Einstein was twenty-six in 1905, no one expects such parallels to be perfect.

While these parallels cannot be denied, upon closer inspection we can also see differences—much more significant than the slight disparity in age—between the activities of the two men during their *anni mirabiles* and in the immediate consequences of their work. The first striking difference is the one between their life situations: rejected by the academic community after graduation from the Swiss Polytechnical School in 1900, by 1905 Einstein was already a married man and an active father of a one-year-old son, obliged to fulfill the demanding responsibilities of a full-time job at the Swiss Patent Office. Newton never married (there is speculation that he died a virgin), and he had just taken his bachelor's degree but was still what we would call a graduate student in 1666. Indeed, he had been temporarily freed of even his academic responsibilities by the closure of Cambridge University after outbreaks of the plague.

Next we may note the difference in their scientific standing. Newton had published nothing by 1666, while Einstein already had published five respectable if not extraordinary papers in the prestigious *Annalen der Physik*. Thus, if 1666 marks the year when Newton's genius caught fire and he embarked on independent research, 1905 marks the year when Einstein's already matured talents manifested themselves to the world in a burst of creativity, a series of epoch-making works, all of which were published by the *Annalen* either in that year or the next. None of Newton's activities in 1666 found their way into print until much later: "The first blossoms of his genius flowered in private, observed

6

silently by his own eyes alone in the years 1664 to 1666, his *anni mirabiles*."[6] The reasons for Newton's evident lack of a need for recognition—indeed, his pronounced reluctance to share his ideas with others, as his major works had to be pried from his hands by others—have long been the topic of psychological, even psychopathological, speculation.

It took a few years—an agonizingly long time for a young man eager for recognition (see p. 115 below)—for Einstein's achievements to be fully acknowledged by the physics community. But the process started almost immediately in 1905; by 1909 Einstein had been called to a chair of theoretical physics created for him at the University of Zurich, and he was invited to lecture at the annual meeting of the assembled German-speaking scientific community.

Thus, if 1905 marks the beginning of the emergence of Einstein as a leading figure in the physics community, Newton remained in self-imposed obscurity well after 1666. Only in 1669, when at the urging of friends he allowed the limited circulation of a mathematical manuscript divulging some parts of the calculus he had developed, did "Newton's anonymity begin to dissolve."[7]

Another striking difference between the two is in their mathematical talents. Newton manifested his mathematical creativity from the outset. "In roughly a year [1664], without the benefit of instruction, he mastered the entire achievement of seventeenth-century analysis and began to break new ground. . . . The fact that he was unknown does not alter the fact that the young man not yet twenty-four, without benefit of formal instruction, had become the leading mathematician of Europe."[8]

Newton was thus able to create the mathematics necessary to develop his ideas about mechanics and gravitation.

7

Einstein, while an able pupil and practitioner, was never really creative in mathematics. Writing about his student years, Einstein said:

> The fact that I neglected mathematics to a certain extent had its cause not merely in my stronger interest in the natural sciences than in mathematics but also in the following peculiar experience. I saw that mathematics was split up into numerous specialties, each of which could easily absorb the short lifetime granted to us. Consequently, I saw myself in the position of Buridan's ass, which was unable to decide upon any particular bundle of hay. Presumably this was because my intuition was not strong enough in the field of mathematics to differentiate clearly the fundamentally important, that which is really basic, from the rest of the more or less dispensable erudition. Also, my interest in the study of nature was no doubt stronger; and it was not clear to me as a young student that access to a more profound knowledge of the more basic principles of physics depends on the most intricate mathematical methods. This dawned upon me only gradually after years of independent scientific work.[9]

Fortunately, for his works of 1905 he needed no more mathematics than he had been taught at school. Even so, it was left to Henri Poincaré, Hermann Minkowski, and Arnold Sommerfeld to give the special theory of relativity its most appropriate mathematical formulation.

When a really crucial need for new mathematics manifested itself in the course of his work on the general theory of relativity, Einstein had to make do with the tensor calculus as developed by Gregorio Ricci-Curbastro and Tullio Levi-Civita and presented to Einstein by his friend and colleague, Marcel Grossmann. This was based on Riemannian geometry, which lacked the concepts of parallel displacement and affine connection that would have so facilitated Einstein's work. But he was incapable of filling this mathematical lacuna, a task that was accomplished by Levi-Civita and Hermann Weyl only after the completion of the general theory.

Returning to Newton: in some respects he was right to hesitate about publication in 1666. "When 1666 closed, Newton was not in command of the results that have made his reputation deathless, not in mathematics, not in mechanics, not in optics. What he had done in all three was to lay foundations, some more extensive than others, on which he could build with assurance, but nothing was complete at the end of 1666, and most were not even close to complete."[10]

His work on the method of fluxions (as he called the calculus), even if incomplete, was worthy of publication and would have been of great service to contemporary mathematicians had it been available to them. His work in physics was far less advanced. His experiments on the theory of colors were interrupted by the closing of the university, and after his return to Cambridge in 1667 he spent a decade pursuing his optical investigations. Nevertheless, a more outgoing man might have published a preliminary account of his theory of colors in 1666. But in the case of gravitation, after carefully reviewing the evidence bearing on Newton's work on this subject through 1666, the physicist Leon Rosenfeld concluded that "it will be clear to every scientist that Newton at this stage had opened up for himself an exciting prospect, but had nothing fit to be published."[11] It is also clear that, in thinking about mechanics, he had not yet arrived at a clear concept of force—an essential prerequisite for the development of what we now call Newtonian mechanics. He had given "a new definition of force in which a body was treated as the passive subject of external forces impressed upon it instead of the active vehicle of forces impinging on others." But: "More than twenty years of patient if intermittent thought would in the end elicit his whole dynamics from this initial insight."[12]

To sum up, in the case of Newton, in 1666 we have a student, working at his leisure, a mature genius in mathematics, but whose work in physics, however genial, was still in its formative stages. In the case of Einstein, in 1905 we have a man raising a family and pursuing a practical career, forced to fit physics into the interstices of an already-full life, yet already a master of theoretical physics ready to demonstrate that mastery to the world.

II

Newton's great legacy was his advancement of what at the time was called the mechanical philosophy and later came to be called the mechanical worldview. In physics, it was embodied in the so-called central force program: matter was assumed to be made up of particles of different species, referred to as "molecules." Two such molecules exerted various forces on each other: gravitational, electrical, magnetic, capillary, etc. These forces—attractive or repulsive—were assumed to be central, that is, to act in the direction of the line connecting the two particles, and to obey appropriate laws (such as the inverse square law for the gravitational and electrostatic forces), which depended on the distance between them. All physical phenomena were assumed to be explicable on the basis of Newton's three laws of motion applied to molecules acted upon by such central forces.

The central force program was shaken around the middle of the nineteenth century when it appeared that, in order to explain electromagnetic interactions between moving charged molecules, velocity- and acceleration-dependent forces had to be assumed. But it received the coup de grâce when Michael Faraday and James Clerk Maxwell's concept of the electromagnetic field began to prevail. According to the field point of view, two charged particles do not interact directly: each charge creates fields in the space surrounding it, and it is these fields which exert forces on the other charge. At first, these electric and magnetic fields were conceived of as states of a mechanical medium, the electromagnetic ether; these states were assumed ultimately to be explainable on the basis of mechanical models of that ether. Meanwhile, Maxwell's equations gave a complete description of the possible states of the electric and magnetic fields at all points of space and how they change over time. By the turn of the century, the search for mechanical explanations of the ether had been largely abandoned in favor of Hendrik Antoon Lorentz's viewpoint, frankly dualistic: the electric and magnetic fields were accepted as fundamental states of the ether, governed by Maxwell's equations but not in need of further explanation. Charged particles, which Lorentz called electrons (others continued to call them molecules or ions), obeyed Newton's mechanical laws of motion under the influence of forces that include the electric and magnetic forces exerted by the ether; and in turn the charged particles created these fields by their presence in and motion through the ether.

I call Lorentz's outlook dualistic because he accepted the mechanical worldview as applied to his electrons but regarded the ether with its electric and magnetic fields as an additional, independent element of reality, not mechanically explicable. To those brought up on the doctrine of the essential unity of nature, especially popular in Germany since the time of Alexander von Humboldt, such a dualism was uncomfortable if not intolerable.

Indeed, it was not long before Wilhelm Wien and others suggested another possibility: perhaps the electromagnetic field is the really fundamental entity, and the behavior of matter depends entirely on its electromagnetic properties. Instead of explaining the behavior of electromagnetic fields in terms of a mechanical model of the ether, this electromagnetic worldview hoped to explain the mechanical properties of matter in terms of electric and magnetic fields. Even Lorentz flirted with this possibility, though he never fully adopted it.

The mechanical worldview did not simply disappear with the advent of Maxwell's electrodynamics. The last third of the nineteenth century saw a remarkable new triumph of the mechanical program. On the basis of the application of statistical methods to large assemblies of molecules (Avogadro's number, about 6.3×10^{23} molecules per mole of any substance, here gives the measure of largeness), Maxwell and Ludwig Boltzmann succeeded in giving a mechanical foundation to the laws of thermodynamics and started the program of explaining the bulk properties of matter in terms of kinetic-molecular theories of the gaseous, liquid, and solid states.

III

Thus, as a student Einstein had to master both the traditional mechanical viewpoint, particularly its application to the atomistic picture of matter, as well as Maxwell's new field-theoretical approach to electromagnetism, particularly in Lorentz's version. He was also confronted with a number of new phenomena, such as black-body radiation and the photoelectric effect, which stubbornly resisted all attempts to fit them into either the old mechanical or the new electromagnetic worldview—or any combination of the two. From this perspective, his five epoch-making papers of 1905 may be divided into three categories. The first two categories concern extensions and modifications of the two physical theories that dominated physics at the end of the nineteenth century: classical mechanics and Maxwell's electrodynamics.

1. His two papers on molecular dimensions and Brownian motion, papers 1 and 2 in this volume, are efforts to extend and perfect the classical-mechanical approach, especially its kinetic-molecular implications.

2. His two papers on the theory of special relativity, papers 3 and 4, are efforts to extend and perfect Maxwell's theory by modifying the foundations of classical mechanics in order to remove the apparent contradiction between mechanics and electrodynamics.

In these four papers, Einstein proved himself a master of what we today call classical physics, the inheritor and continuer of the tradition that started with Galileo Galilei and Newton and ended with Faraday, Maxwell, and Boltzmann, to name but a few of the most outstanding representatives of this tradition. Revolutionary as they then appeared to his

13

contemporaries, the new insights into the nature of space, time, and motion necessary to develop the special theory of relativity are now seen as the climax and culmination of that classical tradition.

3. His work on the light quantum hypotheses, paper 5, is the only one that he himself regarded as truly radical. In the first letter cited on p. 5 above, he wrote that this paper "deals with radiation and the energetic properties of light and is very revolutionary."[13] In it, he demonstrated the limited ability of both classical mechanics and Maxwell's electromagnetic theory to explain the properties of electromagnetic radiation, and introduced the hypothesis that light has a granular structure in order to explain novel phenomena such as the photoelectric effect, which cannot be explained on the basis of classical physics. Here and subsequently, Einstein, master of the classical tradition, proved to be its most severe and consistent critic and a pioneer in the search to find a new unified foundation for all of physics.

IV

The papers are presented in this volume in the order suggested by the three categories mentioned above, roughly the order of their distance from classical physics; but the reader should feel no compulsion to read them in that order. A good case can be made for the chronological order, for jumping immediately to the papers on special relativity and quantum theory—or for simply dipping into the volume as one's interest or fancy dictates.

In the body of this volume, the reader will find detailed discussions of each of these five papers drawn from the thematic introductory essays in volume 2 of *The Collected*

Papers of Albert Einstein. Here I shall give an overview of Einstein's work up to and including 1905 in each of the three categories.

1. Efforts to Extend and Perfect the Classical-Mechanical Tradition

As recently discovered letters show, by the turn of the century Einstein was already occupied with the problems that were to take him beyond classical physics. Yet all of his papers published before 1905 treat topics that fall within the framework of Newtonian mechanics and its applications to the kinetic-molecular theory of matter. In his first two papers, published in 1901 and 1902, Einstein attempted to explain several apparently quite different phenomena occurring in liquids and solutions on the basis of a single simple hypothesis about the nature of the central force between molecules, and how it varies with their chemical composition. Einstein hoped that his work might help to settle the status of a long-standing (and now discarded) conjecture about a common basis for molecular and gravitational forces—one indication of his strong ambition from the outset to contribute to the theoretical unification of all the apparently disparate phenomena of physics. In 1901 he wrote: "It is a wonderful feeling to realize the unity of a complex of phenomena which, to immediate sensory perception, appear to be totally separate things."[14] Much later, looking back over his life, he wrote: "The real goal of my research has always been the simplification and unification of the system of theoretical physics."[15]

As mentioned on p. 12, another great project of nineteenth-century physics was the attempt to show that the empirically well-verified laws of thermodynamics could be explained theoretically on the basis of an atomistic model of matter. Maxwell and Boltzmann were pioneers in this effort, and Einstein saw himself as continuing and perfecting their work.

Einstein made extensive use of thermodynamical arguments in his first two papers; indeed, thermodynamics plays an important role in all of his early work. The second paper raises a question about the relation between the thermodynamic and kinetic-molecular approaches to thermal phenomena that he answered in his next paper. This is the first of three, published between 1902 and 1904, devoted to the atomistic foundations of thermodynamics. His aim was to formulate the minimal atomistic assumptions about a mechanical system needed to derive the basic concepts and principles of thermodynamics. Presumably because he derived it from such general assumptions, he regarded the second law of thermodynamics as a "necessary consequence of the mechanical worldview."[16] He also derived an equation for the mean square energy fluctuations of a system in thermal equilibrium. In spite of its mechanical origins, this formula involves only thermodynamical quantities, and Einstein boldly proceeded to apply the equation to an apparently nonmechanical system: black-body radiation (his first mention of it in print), that is, electromagnetic radiation in thermal equilibrium with matter. Black-body radiation was the only system for which it was clear to him that energy fluctuations should be physically significant on an observable length scale, and his calculations proved consistent with the known properties of that radiation. This calculation suggests that Einstein may already have had in mind an attempt to treat black-body radiation as if it were a mechanical system—the basis of his "very revolutionary" light quantum hypothesis of 1905.

In paper 1 of this volume, his doctoral dissertation, Einstein used methods based on classical hydrodynamics and diffusion theory to show that measurement of a fluid's viscosity with and without the presence of a dissolved substance can be used to obtain an estimate of Avogadro's number (see p. 12) and the size of the molecules of the dissolved substance. Paper 2, the so-called Brownian-motion paper, also extends the scope of applicability of classical mechanical concepts. Einstein noted that, if the kinetic-molecular theory of heat is correct, the laws of thermodynamics cannot be universally valid, since fluctuations must give rise to microscopic but visible violations of the second law when one considers particles sufficiently large for their motion to be observable in a microscope if suspended in a liquid. Indeed, as Einstein showed, such fluctuations explain the well-known Brownian motion of microscopic particles suspended in a liquid. He regarded his work as establishing the limits of validity within which thermodynamics could be applied with complete confidence.

2. Efforts to Extend and Perfect Maxwell's Electrodynamics and Modify Classical Mechanics to Cohere with It

Well before 1905, Einstein apparently was aware of a number of experiments suggesting that the mechanical principle of relativity—the equivalence of all inertial frames of reference for the description of any mechanical phenomena—should be extended from mechanical to optical and electromagnetic phenomena. However, such an extension was in conflict with what he regarded as the best current electro-dynamical theory, Lorentz's electron theory, which grants a privileged status to one inertial frame: the ether rest frame (see p. 11).

In papers 3 and 4 in this volume, Einstein succeeded in resolving this conflict through a critical analysis of the kine-matical foundations of physics, the theory of space and time, which underlies mechanics, electrodynamics, and indeed (although no others were known at the time) any other dynamical theory. After a profound critical study of the concept of simultaneity of distant events, Einstein realized that the principle of relativity could be made compati-ble with Maxwell's equations if one abandoned Newtonian absolute time in favor of a new absolute: the speed of light, the same in all inertial frames. As a consequence, the Newtonian-Galilean laws of transformation between the space and time coordinates of different inertial frames must be replaced by a set of transformations, now called the Lorentz transformations.[17] Since these transformations are kinematical in nature, any acceptable physical theory must be invariant under the group of such transformations. Maxwell's equations, suitably reinterpreted after eliminat-ing the concept of the ether, meet this requirement; but Newton's equations of motion needed revision.

Einstein's work on the theory of relativity provides an example of his ability to move forward amid para-dox and contradiction. He employs one theory—Maxwell's electrodynamics—to find the limits of validity of another—Newtonian mechanics—even though he was already aware of the limited validity of the former (see pp. 20–22 below).

One of the major accomplishments of Einstein's approach, which his contemporaries found difficult to apprehend, is that relativistic kinematics is independent of the theories that impelled its formulation. He had not only formulated a coherent kinematical basis for both mechanics and electrodynamics, but (leaving aside the problem of gravitation) for any new physical concepts that might be introduced. Indeed, developments in physics over almost a century have not shaken these kinematical foundations. To use terms that he employed later, Einstein had created a theory of principle, rather than a constructive theory.[18] At the time he expressed the distinction in these words: "One is in no way dealing here . . . with a 'system' in which the individual laws would implicitly be contained and could be found merely by deduction therefrom, but only with a principle that (in a way similar to the second law of thermodynamics) permits the reduction of certain laws to others."[19] The principles of such a theory, of which thermodynamics is his prime example, are generalizations drawn from a large amount of empirical data that they summarize and generalize without purporting to explain. In contrast, constructive theories, such as the kinetic theory of gases, do purport to explain certain phenomena on the basis of hypothetical entities, such as atoms in motion, introduced precisely to provide such explanations.

It is well known that important elements of Einstein's distinction between principle and constructive theories are found in Poincaré's writings. Two lesser-known sources that may have influenced Einstein's emphasis on the role of principles in physics are the writings of Julius Violle and Alfred Kleiner, which he is also known to have read.

In spite of the merits of the theory of relativity, however, Einstein felt that it was no substitute for a constructive theory: "A physical theory can be satisfactory only if its structures are composed of elementary foundations. The theory of relativity is just as little ultimately satisfactory as, for example, classical thermodynamics was before Boltzmann had interpreted the entropy as probability."[20]

3. Demonstrations of the Limited Validity of Both Classical Mechanics and Maxwell's Electromagnetic Theory, and Attempts to Comprehend Phenomena That Cannot Be Explained by These Theories

Einstein's efforts to perfect classical mechanics and Maxwell's electrodynamics, and to make both theories compatible, may still be regarded as extensions, in the broadest sense, of the classical approach to physics. However original his contributions in these areas may have been, however revolutionary his conclusions about space and time appeared to his contemporaries, however fruitful his work proved to be for the exploration of new areas of physics, he was still engaged in drawing the ultimate consequences from conceptual structures that were well established by the end of the nineteenth century. What is unique about his stance during the first decade of this century is his unwavering conviction that classical mechanical concepts and those of Maxwell's electrodynamics—as well as any mere modification or supplementation of the two—are incapable of explaining a growing list of newly discovered phenomena involving the behavior and interactions of matter and radiation. Einstein constantly reminded his colleagues of the need to introduce radically new concepts to explain the structure of both matter and radiation. He himself introduced some of these new concepts, notably the light quantum hypothesis, although he remained unable to integrate them into a coherent physical theory.

Paper 5, Einstein's first paper on the quantum hypothesis, is a striking example of his style, mingling critique of old concepts with the search for new ones. It opens by demonstrating that the equipartition theorem,[21] together with Maxwell's equations, leads to a definite formula for the black-body radiation spectrum, now known as the Rayleigh-Jeans distribution. This distribution, which at low frequencies matches the empirically validated Planck distribution, cannot possibly hold at high frequencies, since it implies a divergent total energy. (He soon gave a similar demonstration, also based on the equipartition theorem, that classical mechanics cannot explain the thermal or optical properties of a solid, modeled as a lattice of atomic or ionic oscillators.)

Einstein next investigated this high-frequency region, where the classically derived distribution breaks down most dramatically. In this region, called the Wien limit, he showed that the entropy of monochromatic radiation with a fixed temperature depends on its volume in exactly the same way as does the entropy of an ordinary gas composed of statistically independent particles. In short, monochromatic radiation in the Wien limit behaves thermodynamically as if it were composed of statistically independent quanta of energy. To obtain this result, Einstein had to assume each quantum has an energy proportional to its frequency. Emboldened by this result, he took the final step, proposing his "very revolutionary" hypothesis that matter and radiation can interact only through the exchange of such energy quanta. He demonstrated that this hypothesis explains a number of apparently disparate phenomena, notably the photoelectric effect; it was this work that was cited by the Nobel Prize committee in 1921.

In 1905 Einstein did not use Planck's full distribution law. The following year he showed that Planck's derivation of this law implicitly depends on the assumption that the energy of charged oscillators can only be an integral multiple of the quantum of energy, and hence these oscillators can only exchange energy with the radiation field by means of such quanta. In 1907, Einstein argued that uncharged oscillators should be similarly quantized, thereby explaining both the success of the DuLong-Petit law for most solids at ordinary temperatures and the anomalously low values of the specific heats of certain substances. He related the temperature at which departures from the DuLong-Petit law (see p. 175) become significant—now called the Einstein temperature— to the fundamental frequency of the atomic oscillators, and hence to the optical absorption spectrum of a solid.

In spite of his conviction of its fundamental inadequacy, Einstein continued to utilize still-reliable aspects of classical mechanics with remarkable skill to explore the structure of electromagnetic radiation. In 1909 he applied his theory of Brownian motion to a two-sided mirror immersed in thermal radiation. He showed that the mirror would be unable to carry out such a Brownian motion indefinitely if the fluctuations of the radiation pressure on its surfaces were due solely to the effects of random waves, as predicted by Maxwell's theory. Only the existence of an additional term, corresponding to pressure fluctuations due to the impact of random particles on the mirror, guarantees its continued Brownian motion. Einstein showed that both wave and particle energy fluctuation terms are consequences of Planck's distribution law for black-body radiation. He regarded this result as his strongest argument for ascribing physical reality to light quanta.

Einstein was far from considering his work on the quantum hypothesis as constituting a satisfactory theory of radiation or matter. As noted on p. 19, he emphasized that a physical theory is satisfactory only "if its structures are composed of *elementary* foundations," adding "that we are still far from having satisfactory elementary foundations for electrical and mechanical processes."[22] Einstein felt that he had not achieved a real understanding of quantum phenomena because (in contrast to his satisfactory interpretation of Boltzmann's constant as setting the scale of statistical fluctuations) he had been unable to interpret Planck's constant "in an intuitive way."[23] The quantum of electric charge also remained "a stranger" to theory.[24] He was convinced that a satisfactory theory of matter and radiation must construct these quanta of electricity and of radiation, not simply postulate them.

As a theory of principle (see above), the theory of relativity provides important guidelines in the search for such a satisfactory theory. Einstein anticipated the ultimate construction of "a complete worldview that is in accord with the principle of relativity."[25] In the meantime, the theory offered clues to the construction of such a worldview. One clue concerns the structure of electromagnetic radiation. Not only is the theory compatible with an emission theory of radiation, since it implies that the velocity of light is always the same relative to its source; the theory also requires that radiation transfer mass between an emitter and an absorber, reinforcing Einstein's light quantum hypothesis that radiation manifests a particulate structure under certain circumstances. He maintained that "the next phase in the development of theoretical physics will bring us a theory of light, which may be regarded as a sort of fusion of the undulatory and emission theories of light."[26] Other principles that Einstein regarded as reliable guides in the search for an understanding of quantum phenomena are conservation of energy and Boltzmann's principle.

Einstein anticipated that "the same theoretical modification that leads to the elementary quantum [of charge] will also lead to the quantum structure of radiation as a consequence."[27] In 1909 he made his first attempt to find a field theory that would explain both the structure of matter (the electron) and of radiation (the light quantum). After investigating relativistically invariant, non-linear generalizations of Maxwell's equations, he wrote: "I have not succeeded . . . in finding a system of equations that I could see was suited to the construction of the elementary quantum of electricity and the light quantum. The manifold of possibilities does not seem to be so large, however, that one need draw back in fright from the task."[28] This attempt may be regarded as the forerunner of his later, almost forty-year-long search for a unified field theory of electromagnetism, gravitation, and matter.

In 1907, Einstein's attempt to incorporate gravitation into the theory of relativity led him to recognize a new formal principle, the principle of equivalence, which he interpreted as demonstrating the need to generalize the relativity principle (which he now began to call the *special* relativity principle) if gravitation is to be included in its scope. He found that, when gravitational effects are taken into account, it is impossible to maintain the privileged role that inertial frames of reference and Lorentz transformations play in the original relativity theory. He started the search for a group of transformations wider than the Lorentz group, under which the laws of physics remain invariant when gravitation is taken into account. This search, which lasted until the end of 1915, culminated in what Einstein considered his greatest scientific achievement: the general theory of relativity—but that is another story, which I cannot tell here.

Nor can I do more than allude to the many ways in which Einstein's work on the special theory of relativity and the quantum theory have inspired and guided not only many of the revolutionary transformations of our picture of the physical world during the twentieth century, but—through their influence on technological development—have contributed to equally revolutionary transformations in our way of life. One cannot mention quantum optics or quantum field theory, to name only a couple of theoretical advances; nor masers and lasers, klystrons and synchrotrons—nor atomic and hydrogen bombs, to name only a few of the multitude of inventions that have changed our world for good or ill, without invoking the heritage of Einstein's miraculous year.

EDITORIAL NOTES

[1]The phrase *anni mirabiles* (years of wonders) has been applied with more accuracy to the years 1664–1666 by Newton's biographer Richard Westfall in *Never at Rest/A Biography of Isaac Newton* (Cambridge, U.K.: Cambridge University Press, 1980; paperback edition, 1983), p. 140. This book may be consulted for generally reliable biographical information about Newton's life.

[2]I. Bernard Cohen, *Introduction to Newton's 'Principia'* (Cambridge, Mass.: Harvard University Press, 1971), p. 291.

[3]See, for example, Albrecht Fölsing, *Albert Einstein/A Biography*, tr. by Ewald Osers (New York: Viking, 1997), p. 121: "Never before and never since has a single person enriched science by so much in such a short time as Einstein did in his *annus mirabilis*." This book may be consulted for generally reliable biographical information about Einstein, but its scientific explanations should be treated with caution. For an account of Einstein's scientific work organized biographically, see Abraham Pais,

'*Subtle is the Lord . . .* ': *The Science and the Life of Albert Einstein* (Oxford: Clarendon Press; New York: Oxford University Press, 1982).

[4]Einstein to Conrad Habicht, 18 or 25 May 1905, *The Collected Papers of Albert Einstein* (Princeton, N.J.: Princeton University Press, 1987–), cited hereafter as *Collected Papers*, vol. 5 (1993), doc. 27, p. 31. Translation from Anna Beck, tr., *The Collected Papers of Albert Einstein: English Translation* (Princeton University Press, 1987–), cited hereafter as *English Translation*, vol. 5 (1995), p. 20; translation modified.

[5]Einstein to Conrad Habicht, 30 June–22 September 1905, *Collected Papers*, vol. 5, doc. 28, p. 33; *English Translation*, p. 21; translation modified. Forty years later, when the explosion of the first atomic bombs brought the equivalence between mass and energy forcefully to the world's attention, Einstein might have wondered just what sort of trick the Lord had played on him.

[6]Westfall, *Never at Rest*, p. 140.

[7]Ibid., p. 205.

[8]Ibid., pp. 100, 137.

[9]Albert Einstein, *Autobiographical Notes*, Paul Arthur Schilpp, ed. and trans. (LaSalle, Ill.: Open Court, 1979), p. 15.

[10]Westfall, *Never at Rest*, p. 174.

[11]"Newton and the Law of Gravitation," *Arch. Hist. Exact Sci.* 2 (1965): 365–386, reprinted in Robert S. Cohen and John J. Stachel, eds., *Selected Papers of Leon Rosenfeld* (Dordrecht/Boston: Reidel, 1979), p. 65.

[12]Citation from Westfall, *Never at Rest*, p. 146.

[13]Einstein to Conrad Habicht, May 1905, *Collected Papers*, vol. 5, doc. 27, p. 31.

[14]Einstein to Marcel Grossmann, 14 April 1901, *Collected Papers*, vol. 1, doc. 100, p. 290.

[15]From the response to a questionnaire submitted to Einstein in 1932. See Helen Dukas and Banesh Hoffmann, *Albert Einstein: The Human Side* (Princeton, N.J.: Princeton University Press, 1979), p. 11 for the English translation, p. 122 for the German text.

[16]Einstein, "Kinetic Theory of Thermal Equilibrium and of the Second Law of Thermodynamics," in *Collected Papers*, vol. 2, doc. 3, p. 72 (p. 432 of the 1902 original).

[17]Lorentz had introduced such a set of transformations, and Henri Poincaré had so named them; but the kinematical interpretation that Einstein gave to them is quite different.

[18]For the distinction between theories of principle and constructive theories, see Albert Einstein, "Time, Space and Gravitation," *The Times* (London), 28 November 1919, p. 13; reprinted as "What Is the Theory of

INTRODUCTION

Relativity?" in *Ideas and Opinions* (New York: Crown, 1954), pp. 227–232. He later reminisced about the origins of the theory: "Gradually I despaired of the possibility of discovering the true laws by means of constructive efforts based on known facts. The longer and the more desperately I tried, the more I came to the conviction that only the discovery of a universal formal principle could lead us to assured results. The example I saw before me was thermodynamics" (*Autobiographical Notes*, p. 48; translation, p. 49). For several years after 1905, Einstein referred to the "relativity principle" rather than to the "theory of relativity."

[19] Einstein, "Comments on the Note of Mr. Paul Ehrenfest: 'The Translatory Motion of Deformable Electrons and the Area Law,' " in *Collected Papers*, vol. 2, doc. 44, p. 411 (p. 207 of the 1907 original).

[20] Einstein to Arnold Sommerfeld, 14 January 1908, *Collected Papers*, vol. 5, doc. 73, pp. 86–88. A decade later, Einstein elaborated this idea: "When we say that we have succeeded in understanding a group of natural processes, we always mean by this that a constructive theory has been found, which embraces the processes in question" (from "Time, Space and Gravitation").

[21] This is a result of classical statistical mechanics, according to which each degree of freedom of a mechanical system in thermal equilibrium receives, on the average, the same share of the total energy of the system.

[22] Einstein to Arnold Sommerfeld, 14 January 1908, *Collected Papers*, vol. 5, doc. 73, p. 87.

[23] Ibid.

[24] See Einstein, "On the Present Status of the Radiation Problem," in *Collected Papers*, vol. 2, doc. 56, p. 549 (p. 192 of the 1909 original).

[25] Einstein, "On the Inertia of Energy Required by the Relativity Principle," in *Collected Papers*, vol. 2, doc. 45, pp. 414–415 (p. 372 of the 1907 original).

[26] Einstein, "On the Development of Our Views Concerning the Nature and Constitution of Radiation," in *Collected Papers*, vol. 2, doc. 60, pp. 564–565 (pp. 482–483 of the 1909 original).

[27] Einstein, "On the Present Status of the Radiation Problem," in *Collected Papers*, vol. 2, doc. 56, pp. 549–550 (pp. 192–193 of the 1909 original).

[28] Ibid., p. 550 (p. 193 of the 1909 original). This attempt at a field theory seems to represent Einstein's first step toward a field ontology.

Part One

✦

Einstein's Dissertation on
the Determination of
Molecular Dimensions

Lecture hall in the Physics Building, Eidgenössische
Technische Hochschule, Zurich, 1905. (Courtesy of ETH)

Einstein submitted a dissertation to the University of Zurich in 1901, about a year after graduation from the Eidgenössische Technische Hochschule (ETH), but withdrew it early in 1902. In a successful second attempt three years later, he combined the techniques of classical hydrodynamics with those of the theory of diffusion to create a new method for the determination of molecular sizes and of Avogadro's number, a method he applied to solute sugar molecules. The dissertation was completed on 30 April 1905 and submitted to the University of Zurich on 20 July. On 19 August 1905, shortly after the thesis was accepted, the *Annalen der Physik* received a slightly different version for publication.

By 1905, several methods for the experimental determination of molecular dimensions were available. Although estimates of upper bounds for the sizes of microscopic constituents of matter had been discussed for a long time, the first reliable methods for determining molecular sizes were developed in the second half of the nineteenth century, based on the kinetic theory of gases. The study of phenomena as diverse as contact electricity in metals, the dispersion of light, and black-body radiation yielded new approaches to the problem of molecular dimensions. Most of the methods available by the turn of the century gave values for the size of molecules and for Avogadro's number that are in more or less satisfactory agreement with each other.

Although Einstein claimed that the method in his dissertation is the first to use phenomena in fluids in the determination of molecular dimensions, the behavior of liquids plays a role in various earlier methods. For example, the comparison of densities in the liquid and gaseous states is an important part of Loschmidt's method, based on the kinetic theory of gases. A method that depends entirely on the physics of liquids was developed as early as 1816 by Thomas Young. Young's study of surface tension in liquids led to an estimate of the range of molecular forces, and capillary phenomena were used later in several different ways to determine molecular sizes.

A kinetic theory of liquids, comparable to the kinetic theory of gases, was not available, and the methods for deriving molecular volumes exclusively from the properties of liquids did not give very precise results. Einstein's method, on the other hand, yields values comparable in precision to those provided by the kinetic theory of gases. While methods based on capillarity presuppose the existence of molecular forces, Einstein's central assumption is the validity of using classical hydrodynamics to calculate the effect of solute molecules, treated as rigid spheres, on the viscosity of the solvent in a dilute solution.

Einstein's method is well suited to determine the size of solute molecules that are large compared to those of the solvent. In 1905 William Sutherland published a new method for determining the masses of large molecules that shares important elements with Einstein's. Both methods make use of the molecular theory of diffusion that Nernst developed on the basis of van't Hoff's analogy between solutions and gases, and of Stokes's law of hydrodynamical friction.

Sutherland was interested in the masses of large molecules because of the role they play in the chemical analysis of organic substances such as albumin. In developing a new method for the determination of molecular dimensions, Einstein was concerned with several other problems on different levels of generality. An outstanding current problem of the theory of solutions was whether molecules of the solvent are attached to the molecules or ions of the solute. Einstein's dissertation contributed to the solution of this problem. He recalled in a letter to Jean Perrin in November 1909: "At the time I used the viscosity of the solution to determine the volume of sugar dissolved in water because in this way I hoped to take into account the volume of any *attached* water molecules." The results obtained in his dissertation indicate that such an attachment does occur.

Einstein's concerns extended beyond this particular question to more general problems of the foundations of the theory of radiation and the existence of atoms. He later emphasized in the same letter: "A precise determination of the size of molecules seems to me of the highest importance because Planck's radiation formula can be tested more precisely through such a determination than through measurements on radiation."

The dissertation also marked the first major success in Einstein's effort to find further evidence for the atomic hypothesis, an effort that culminated in his explanation of Brownian motion. By the end of 1905 he had published three independent methods for determining molecular dimensions, and in the following years he found several more. Of all these methods, the one in his dissertation is most closely related to his earlier studies of physical phenomena in liquids.

EINSTEIN'S efforts to obtain a doctoral degree illuminate some of the institutional constraints on the development of his work on the problem of molecular dimensions. His choice of a theoretical topic for a dissertation at the University of Zurich was quite unusual, both because it was theoretical and because a dissertation theme was customarily assigned by the supervising professor. By 1900, theoretical physics was slowly beginning to achieve recognition as an independent discipline in German-speaking countries, but it was not yet established at either the ETH or the University of Zurich. A beginning had been made at the ETH soon after its founding, with the appointment of a German mathematical physicist, Rudolf Clausius. His departure a decade later may have been hastened by lack of official sympathy for a too-theoretical approach to the training of engineers and secondary-school teachers, the primary task of the school.

Clausius's successor—after the position had been vacant for a number of years—was H. F. Weber, who occupied the chair for Mathematical and Technical Physics from 1875 until his death in 1912. During the last two decades of the nineteenth century, he did original research, mainly in experimental physics and electrotechnology, including work on a number of topics that were important for Einstein's later research, such as black-body radiation, the anomalous low-temperature behavior of specific heats, and the theory of diffusion; but his primary interests were never those of a theoretical physicist. The situation of theoretical physics at the University of Zurich at the turn of the century was hardly better. Four other major Swiss universities either had two full professorships in physics or one full and one nontenured position, while Zurich had only one physics chair, held by the experimentalist Alfred Kleiner.

Since the ETH was not authorized to grant doctoral degrees until 1909, a special arrangement enabled ETH students to obtain doctorates from the University of Zurich. Most dissertations in physics by ETH students were prepared under Weber's supervision, with Kleiner as the second referee. As noted above, almost all physics dissertations prepared at the ETH and the University of Zurich between 1901 and 1905 were on experimental topics suggested to the students by their supervisor or at least closely related to the latter's research interests. The range of topics was quite limited, and generally not at the forefront of experimental research. Thermal and electrical conductivity, and instruments for their measurement, were by far the most prominent subjects. General questions of theoretical physics, such as the properties of the ether or the kinetic theory of gases, occasionally found their way into examination papers, but they were hardly touched upon in dissertations.

In the winter semester of 1900–1901, Einstein intended to work for a degree under Weber. The topic may have been related to thermoelectricity, a field in which Einstein had shown an interest and in which several of Weber's doctoral students did experimental research. After a falling-out with Weber, Einstein turned to Kleiner for advice and comments on his work.

Although Kleiner's research at this time focused on measuring instruments, he did have an interest in foundational questions of physics, and Einstein's discussions with him covered a wide range of topics. Einstein showed his first dissertation to Kleiner before submitting it to the university in November 1901. This dissertation has not survived, and the evidence concerning its contents is somewhat ambiguous. In April 1901 Einstein wrote that he planned to summarize his work on molecular forces, up to that time mainly on liquids; at the end of the year, his future wife Mileva Marić stated that he had submitted a work on molecular forces in gases. Einstein himself wrote that it concerned "a topic in the

35

kinetic theory of gases." There are indications that the dissertation may have discussed Boltzmann's work on gas theory, as well as Drude's work on the electron theory of metals. By February 1902 Einstein had withdrawn the dissertation, possibly at Kleiner's suggestion that he avoid a controversy with Boltzmann. In view of the predominantly experimental character of the physics dissertations submitted to the University of Zurich at the time, lack of experimental confirmation for his theoretical results may have played a role in the decision to withdraw the thesis. In January 1903 · Einstein still expressed interest in molecular forces, but he stated in a letter to Michele Besso that he was giving up his plan to obtain a doctorate, arguing that it would be of little help to him, and that "the whole comedy has become tiresome for me."

Little is known about when Einstein started to work on the dissertation he completed in 1905. By March 1903 some of the central ideas of the 1905 dissertation had already occurred to him. Kleiner, one of the two faculty reviewers of his dissertation, acknowledged in his review that Einstein had chosen the topic himself and pointed out that "the arguments and calculations to be carried out are among the most difficult in hydrodynamics." The other reviewer, Heinrich Burkhardt, Professor of Mathematics at the University of Zurich, added: "The mode of treatment demonstrates fundamental mastery of the relevant mathematical methods." Although Burkhardt checked Einstein's calculations, he overlooked a significant error in them. The only reported criticism of Einstein's dissertation was for being too short. Einstein's biographer Carl Seelig reports: "Einstein later laughingly recounted that his dissertation was at first returned to him by Kleiner with the comment that it was too short. After he had added a single sentence, it was accepted without further comment."

Compared to the other topics of his research at the time, his hydrodynamical method for determining molecular dimensions was a dissertation topic uniquely suited to the empirically oriented Zurich academic environment. In contrast to the Brownian-motion work, for which the experimental techniques needed to extract information from observations were not yet available, Einstein's hydrodynamical method for determining the dimensions of solute molecules enabled him to derive new empirical results from data in standard tables.

LIKE Loschmidt's method based on the kinetic theory of gases, Einstein's method depends on two equations for two unknowns, Avogadro's number N and the molecular radius P. The first of Einstein's equations (see third equation on p. 64) follows from a relation between the coefficients of viscosity of a liquid with and without suspended molecules (k^* and k, respectively),

$$k^* = k(1 + \varphi),$$ (1)

where φ is the fraction of the volume occupied by the solute molecules. This equation, in turn, is derived from a study of the dissipation of energy in the fluid.

Einstein's other fundamental equation follows from an expression for the coefficient of diffusion D of the solute. This expression is obtained from Stokes's law for a sphere of radius P moving in a liquid, and van't Hoff's law for the osmotic pressure:

$$D = \frac{RT}{6\pi k} \cdot \frac{1}{NP},$$ (2)

where R is the gas constant, T the absolute temperature, and N Avogadro's number.

The derivation of eq. (1), technically the most complicated part of Einstein's thesis, presupposes that the motion of the fluid can be described by the hydrodynamical equations for stationary flow of an incompressible homogeneous liquid, even in the presence of solute molecules; that the inertia of these molecules can be neglected; that they do not affect each other's motions; and that they can be treated as rigid spheres moving in the fluid without slipping, under the sole influence of hydrodynamical stresses. The hydrodynamic techniques needed are derived from Kirchhoff's *Vorlesungen über mathematische Physik*, volume 1, *Mechanik* (1897), a book that Einstein first read during his student years.

Eq. (2) follows from the conditions for the dynamical and thermodynamical equilibrium of the fluid. Its derivation requires the identification of the force on a single molecule, which appears in Stokes's law, with the apparent force due to the osmotic pressure. The key to handling this problem is the introduction of fictitious countervailing forces. Einstein had earlier introduced such fictitious forces to counteract thermodynamical effects in proving the applicability to diffusion phenomena of a generalized form of the second law of thermodynamics, and in his papers on statistical physics.

Einstein's derivation of eq. (2) does not involve the theoretical tools he developed in his work on the statistical foundations of thermodynamics; he reserved a more elaborate derivation, using these methods, for his first paper on Brownian motion. Eq. (2) was derived independently, in somewhat more general form, by Sutherland in 1905. To deal with the available empirical data, Sutherland had to allow for a varying coefficient of sliding friction between the diffusing molecule and the solution.

The basic elements of Einstein's method—the use of diffusion theory and the application of hydrodynamical techniques to phenomena involving the atomistic constitution of matter or electricity—can be traced back to his earlier work. Einstein's previous work had touched upon most aspects of the physics of liquids in which their molecular structure is assumed to play a role, such as Laplace's theory of capillarity, Van der Waals's theory of liquids, and Nernst's theory of diffusion and electrolytic conduction.

Before Einstein's dissertation, the application of hydrodynamics to phenomena involving the atomic constitution of matter or electricity was restricted to consideration of the effects of hydrodynamical friction on the motion of ions. Stokes's law was employed in methods for the determination of the elementary charge and played a role in studies of electrolytic conduction. Einstein's interest in the theory of electrolytic conduction may have been decisive for the development of some of the main ideas in his dissertation. This interest may have suggested a study of molecular aggregates in combination with water, as well as some of the techniques used in the dissertation.

In 1903 Einstein and Besso discussed a theory of dissociation that required the assumption of such aggregates, the "hypothesis of ionic hydrates," as Besso called it, claiming that this assumption resolves difficulties with Ostwald's law of dilution. The assumption also opens the way to a simple calculation of the sizes of ions in solution, based on hydrodynamical considerations. In 1902 Sutherland had considered a calculation of the sizes of ions on the basis of Stokes's formula, but rejected it as in disagreement with experimental data. Sutherland did not use the assumption of ionic hydrates, which can avoid such disagreement by permitting ionic sizes to vary with such physical conditions as temperature and concentration. The idea of determining the sizes of ions by means of classical hydrodynamics occurred to Einstein in March 1903, when he proposed in a letter to Besso what appears to be just the calculation that Sutherland had rejected:

> Have you already calculated the absolute magnitude of ions on the assumption that they are spheres and so large that the hydrodynamical equations for viscous fluids are applicable? With our knowledge of the absolute magnitude of the electron [charge] this would be a simple matter indeed. I would have done it myself but lack the reference material and the time; you could also bring in diffusion in order to obtain information about neutral salt molecules in solution.

This passage is remarkable, because both key elements of Einstein's method for the determination of molecular dimensions, the theories of hydrodynamics and diffusion, are already mentioned, although the reference to hydrodynamics probably covers only Stokes's law. While a program very similar to the first of Einstein's proposals to Besso was being pursued at the time by William Robert Bousfield, Einstein's dissertation can be seen to be an elaboration of the second proposal, regarding diffusion and neutral salt molecules. Einstein may thus have been proceeding similarly to Nernst, who first developed his theory of diffusion for the simpler case of nonelectrolytes. The study of sugar solutions could draw upon extensive and relatively precise numerical data on viscosity and the diffusion coefficient, avoiding problems of dissociation and electrical interactions.

THE RESULTS obtained with Einstein's method for the determination of molecular dimensions differed from those obtained by other methods at the time, even when new data taken from Landolt and Bornstein's physical-chemical tables were used to recalculate them. In his papers on Brownian motion, Einstein cited either the value he obtained for Avogadro's number, or a more standard one. Only once, in 1908, did he comment on the uncertainty in the determination of this number. By 1909 Perrin's careful measurements of Brownian motion produced a new value for Avogadro's number, significantly different from the values Einstein obtained from his hydrodynamical method and from Planck's black-body radiation law. For Einstein, this discrepancy was particularly significant in view of what he regarded as the problematic nature of Planck's derivation of the radiation law.

In 1909 Einstein drew Perrin's attention to his hydrodynamical method for determining the size of solute molecules. He emphasized that this method allows one to take into account the volume of any water molecules attached to the solute molecules, and suggested its application to the suspensions studied by Perrin. In the following year, an experimental study of Einstein's formula for the viscosity coefficients (eq. [1] above) was performed in Perrin's laboratory by Jacques Bancelin. Bancelin studied uniform aqueous emulsions of gamboge, prepared with the help of Perrin's method of fractional centrifugation. Bancelin confirmed that the increase in viscosity does not depend on the size of the suspended particles, but only on the fraction of the total volume that they occupy. However, he found a value for the increased viscosity that differs significantly from Einstein's prediction. Bancelin sent a report of his experiments to

Einstein, apparently citing a value of 3.9 for the coefficient of φ in eq. (1), instead of the predicted value of 1.

After an unsuccessful attempt to find an error in his calculations, Einstein wrote to his student and collaborator Ludwig Hopf in January 1911: "I have checked my previous calculations and arguments and found no error in them. You would be doing a great service in this matter if you would carefully recheck my investigation. Either there is an error in the work, or the volume of Perrin's suspended substance in the suspended state is greater than Perrin believes."

Hopf found an error in the derivatives of the velocity components, which occur in the equations for the pressure components in Einstein's dissertation (see pp. 53–54 below). After correction of this error, the coefficient of φ in eq. (1) becomes 2.5.

By mid-January 1911 Einstein had informed Bancelin and Perrin of Hopf's discovery of the error in his calculations. The remaining discrepancy between the corrected factor 2.5 in eq. (1) and Bancelin's experimental value of 3.9 led Einstein to suspect that there might also be an experimental error. He asked Perrin: "Wouldn't it be possible that your mastic particles, like colloids, are in a swollen state? The influence of such a swelling 3.9/2.5 would be of rather slight influence on Brownian motion, so that it might possibly have escaped you."

On 21 January 1911, Einstein submitted his correction for publication. In the *Annalen der Physik* he presented the corrected form of some of the equations in the dissertation and recalculated Avogadro's number. He obtained a value of 6.56×10^{23} per mole, a value that is close to those derived from kinetic theory and Planck's black-body radiation formula.

Bancelin continued his experiments, with results that brought experiment and theory into closer agreement. Four months later, he presented a paper on his viscosity measurements to the French Academy of Sciences, giving a value of 2.9 as the coefficient of φ in eq. (1). Bancelin also recalculated Avogadro's number by extrapolating his results for emulsions to sugar solutions, and found a value of 7.0×10^{23} per mole.

Einstein's dissertation was at first overshadowed by his more spectacular work on Brownian motion, and it required an initiative by Einstein to bring it to the attention of his fellow scientists. But the wide variety of applications of its results ultimately made the dissertation one of his most frequently cited papers.

PAPER 1

✦

A New Determination of
Molecular Dimensions

(*Ph.D. Dissertation, University of Zurich*)

THE EARLIEST determinations of the real sizes of molecules were made possible by the kinetic theory of gases, but thus far the physical phenomena observed in liquids have not helped in ascertaining molecular sizes. No doubt this is because it has not yet been possible to surmount the hurdles that impede the development of a detailed molecular-kinetic theory of liquids. It will be shown in this paper that the size of molecules of substances dissolved in an undissociated dilute solution can be determined from the internal viscosity of the solution and of the pure solvent, and from the diffusion rate of the solute within the solvent provided that the volume of a solute molecule is large compared to the volume of a solvent molecule. This is possible because, with respect to its mobility within the solvent and its effect on the viscosity of the latter, such a molecule will behave approximately like a solid body suspended in a solvent. Thus, in the immediate vicinity of a molecule, one can apply the equations of hydrodynamics to the motion of the solvent in which the liquid is treated as homogeneous and hence its molecular structure need not be taken into consideration. We will choose a sphere as the solid body that shall represent the solute molecules.

1. How a Very Small Sphere Suspended in a Liquid Influences Its Motion

Let us base our discussion on an incompressible homogeneous liquid with a coefficient of viscosity k, whose velocity components u, v, w are given as functions of the coordinates x, y, z and of time. At an arbitrary point x_0, y_0, z_0, let us think of the functions u, v, w as functions of $x - x_0$, $y - y_0$, $z - z_0$ expanded in a Taylor's series, and of a region G around this point so small that within it only the linear terms of this expansion need be considered. As is well known, the motion of the liquid within G can then be regarded as a superposition of three motions:

1. A parallel displacement of all particles of the liquid without a change in their relative positions;
2. A rotation of the liquid without a change in the relative positions of the particles of the liquid;
3. A dilational motion in three mutually perpendicular directions (the principal axes of dilation).

Let us now assume that in region G there is a spherical rigid body whose center lies at the point x_0, y_0, z_0 and whose dimensions are very small compared with those of region G. We further assume that the motion is so slow that the kinetic energy of the sphere as well as that of the liquid can be neglected. We also assume that the velocity components of a surface element of the sphere coincide with the corresponding velocity components of the adjacent liquid particles, i.e., that the contact layer (imagined to be continuous) also displays a coefficient of viscosity that is not infinitesimally small.

It is obvious that the sphere simply takes part in the partial motions 1 and 2 without altering the motion of neighboring particles, since the liquid moves like a rigid body in these partial motions and since we have neglected the effects of inertia.

However, motion 3 does get altered by the presence of the sphere, and our next task will be to investigate the effect of the sphere on this motion of the liquid. If we refer motion 3 to a coordinate system whose axes are parallel to the principal axes of dilation and set

$$x - x_0 = \xi,$$
$$y - y_0 = \eta,$$
$$z - z_0 = \zeta,$$

we can describe the above motion, if the sphere is not present, by the equations

$$\begin{cases} u_0 = A\xi, \\ v_0 = B\eta, \\ w_0 = C\zeta; \end{cases} \tag{1}$$

A, B, C are constants that, because the liquid is incompressible, satisfy the condition

$$A + B + C = 0. \tag{2}$$

If, now, a rigid sphere of radius P is introduced at the point x_0, y_0, z_0, the motion of the liquid around it will change. We will, for convenience, call P "finite," but all the values of

ξ, η, ζ, for which the liquid motion is no longer noticeably altered by the sphere, we will call "infinitely large."

Because of the symmetry of the motion of the liquid being considered, it is clear that during this motion the sphere can perform neither a translation nor a rotation, and we obtain the boundary conditions

$$u = v = w = 0 \text{ when } \rho = P,$$

where

$$\rho = \sqrt{\xi^2 + \eta^2 + \zeta^2} > 0.$$

Here u, v, w denote the velocity components of this motion (changed by the sphere). If we set

$$\begin{aligned} u &= A\xi + u_1, \\ v &= B\eta + v_1, \\ w &= C\zeta + w_1, \end{aligned} \tag{3}$$

the velocities u_1, v_1, w_1 would have to vanish at infinity, since at infinity the motion represented in equations (3) should reduce to that represented by equations (1).

The functions u, v, w have to satisfy the equations of hydrodynamics, including viscosity and neglecting inertia. Thus the following equations will hold:[1]

$$\begin{cases} \dfrac{\delta p}{\delta \xi} = k\Delta u \dfrac{\delta p}{\delta \eta} = k\Delta v \dfrac{\delta p}{\delta \zeta} = \Delta w,^{[1]} \\ \dfrac{\delta u}{\delta \xi} + \dfrac{\delta v}{\delta \eta} + \dfrac{\delta w}{\delta \zeta} = 0, \end{cases} \tag{4}$$

[1] G. Kirchhoff, *Vorlesungen über Mechanik*, 26. Vorl. (*Lectures on Mechanics*, Lecture 26).

where Δ denotes the operator

$$\frac{\delta^2}{\delta\xi^2} + \frac{\delta^2}{\delta\eta^2} + \frac{\delta^2}{\delta\zeta^2}$$

and p the hydrostatic pressure.

Since equations (1) are solutions of equations (4) and the latter are linear, according to (3) the quantities u_1, v_1, w_1 must also satisfy equations (4). I determined u_1, v_1, w_1, and p by a method given in section 4 of the Kirchhoff lectures mentioned above[2] and found

$$\begin{cases} p = -\tfrac{5}{3}kP^3 \left\{ A\dfrac{\delta^2\left[\frac{1}{\rho}\right]}{\delta\xi^2} + B\dfrac{\delta^2\left[\frac{1}{\rho}\right]}{\delta\eta^2} + C\dfrac{\delta^2\left[\frac{1}{\delta}\right]}{\delta\zeta^2} \right\} + \text{const.}, \\[2mm] u = A\xi - \dfrac{5}{3}P^3 A\dfrac{\xi}{\rho^3} - \dfrac{\delta D}{\delta\xi}, \\[2mm] v = B\eta - \dfrac{5}{3}P^3 B\dfrac{\eta}{\rho^3} - \dfrac{\delta D}{\delta\eta}, \\[2mm] w = C\zeta - \dfrac{5}{3}P^3 C\dfrac{\zeta}{\rho^3} - \dfrac{\delta D}{\delta\zeta}, \end{cases} \qquad (5)^{[5]}$$

---

[2] "From equations (4) it follows that $\Delta p = 0$. If we take p in accordance with this condition and determine a function V that satisfies the equation

$$\Delta V = \frac{1}{k}p,$$

then equations (4) are satisfied if one sets

$$u = \frac{\delta V}{\delta\xi} + u', \quad v = \frac{\delta V}{\delta\eta} + v', \quad w = \frac{\delta V}{\delta\zeta} + w',$$

and chooses u', v', w' such that $\Delta u' = 0$, $\Delta v' = 0$, $\Delta w' = 0$, and

$$\frac{\delta u'}{\delta\xi} + \frac{\delta v'}{\delta\eta} + \frac{\delta w'}{\delta\zeta} = -\frac{1}{k}p."$$

Now, if one sets

$$\frac{p}{k} = 2c\frac{\delta^2\frac{1}{p}}{\delta\xi^3}, \; ^{[2]}$$

49

where

$$
\begin{cases}
D = & A\left\{ \tfrac{5}{6}\, p^3 \dfrac{\delta^2 \rho}{\delta \xi^2} + \tfrac{1}{6} P^5 \dfrac{\delta^2 \left(\tfrac{1}{\rho}\right)}{\delta \xi^2} \right\} \\[4mm]
& + B\left\{ \tfrac{5}{6}\, p^3 \dfrac{\delta^2 \rho}{\delta \eta^2} + \tfrac{1}{6} P^5 \dfrac{\delta^2 \left(\tfrac{1}{\rho}\right)}{\delta \eta^2} \right\} \\[4mm]
& + C\left\{ \tfrac{5}{6} p^3 \dfrac{\delta^2 \rho}{\delta \zeta^2} + \tfrac{1}{6} P^5 \dfrac{\delta^2 \left(\tfrac{1}{\rho}\right)}{\delta \zeta^2} \right\}.
\end{cases} \tag{5a}
$$

It can easily be proved that equations (5) are solutions of equations (4). Since

$$
\Delta \xi = 0, \quad \Delta \frac{1}{\rho} = 0, \quad \Delta \rho = \frac{2}{\rho}
$$

and

$$
\Delta\left(\frac{\xi}{\rho^3}\right) = -\frac{\delta}{\delta \xi}\left\{\Delta\left(\frac{1}{\rho}\right)\right\} = 0,
$$

we get

$$
\begin{aligned}
k\Delta u &= -k\frac{\delta}{\delta \xi}\{\Delta D\} \\
&= -k\frac{\delta}{\delta \xi}\left\{ \frac{5}{3} P^3 A \frac{\delta^2 \frac{1}{\rho}}{\delta \xi^2} + \frac{5}{3} P^3 B \frac{\delta^2 \frac{1}{\rho}}{\delta \eta^2} + \cdots \right\}.
\end{aligned}
$$

---

and, in accordance with this,

$$
V = c\,\frac{\delta^2 \rho}{\delta \xi^3} + b\frac{\delta^2 \frac{1}{\rho}}{\delta \xi^2} + \frac{a}{2}\left[\xi^2 - \frac{\eta^2}{2} - \frac{\zeta^2}{2} \right] \text{[3]}
$$

and

$$
u' = -2c\frac{\delta \frac{1}{\delta}}{\delta \xi}, \quad v' = 0, \quad w' = 0, \text{[4]}
$$

then the constants a, b, c can be determined such that $u = v = w = 0$ for $\rho = P$. By superposing three such solutions, we get the solution given in equations (5) and (5a).

However, according to the first of equations (5), the last of the expressions we obtained is identical to $\frac{\delta n}{\delta \xi}$.[6] In the same way, it can be shown that the second and third of equations (4) are satisfied. Further, we get

$$\frac{\delta u}{\delta \xi} + \frac{\delta v}{\delta \eta} + \frac{\delta w}{\delta \xi} = (A + B + C)$$
$$+ \tfrac{5}{3} P^3 \left\{ A \frac{\delta^2 \left(\frac{1}{\rho}\right)}{\delta \xi^2} + B \frac{\delta^2 \left(\frac{1}{\rho}\right)}{\delta \eta^2} + C \frac{\delta^2 \left(\frac{1}{\rho}\right)}{\delta \zeta^2} \right\} - \Delta D.$$

But since according to equation (5a)

$$\Delta D = \tfrac{5}{3} A P^3 \left\{ A \frac{\delta^2 \left(\frac{1}{\rho}\right)}{\delta \xi^2} + B \frac{\delta^2 \left(\frac{1}{\rho}\right)}{\delta \eta^2} + C \frac{\delta^2 \left(\frac{1}{\rho}\right)}{\delta \zeta^2} \right\},$$

it follows that the last of equations (4) is satisfied as well. As far as the boundary conditions are concerned, at infinitely large ρ our equations for u, v, w reduce to equations (1). By inserting the value of D from equation (5a) into the second of equations (5), we get

$$u = A \xi - \tfrac{5}{2} \frac{P^3}{\rho^6} \xi (A \xi^2 + B \eta^2 + C \zeta^2)^{[7]}$$
$$+ \tfrac{5}{2} \frac{P^5}{\rho^7} \xi (A \xi^2 + B \eta^2 + C \zeta^2) - \frac{P^5}{\rho^5} A \xi.$$

$$(6)$$

We see that u vanishes for $\rho = P$. For reasons of symmetry, the same holds for v and w. We have now demonstrated that equations (5) satisfy equations (4) as well as the boundary conditions of the problem.

It can also be demonstrated that equations (5) are the only solution of equations (4) that is compatible with the boundary conditions of our problem. The proof will only be outlined here. Assume that in a finite region the velocity components u, v, w of a liquid satisfy equations (4). If there existed yet another solution U, V, W for equations (4) in which $U = u$, $V = v$, $W = w$ at the boundaries of the region in question, then $(U - u, V - v, W - w)$ would be a solution for equations (4) in which the velocity components vanish at the boundary. Thus no mechanical work is supplied to the liquid in the region in question. Since we have neglected the kinetic energy of the liquid, it follows that in this volume the work converted to heat is also zero. This leads to the conclusion that in the entire space we must have $u = u_1$, $v = v_1$, $w = w_1$ if the region is at least partly bounded by stationary walls.[8] By passing to the limit, this result can also be extended to a case where the region is infinite, as in the case considered above. One can thus show that the solution found above is the only solution to the problem.

We now draw a sphere of radius R around point x_0, y_0, z_0, with R infinitely large compared to P, and calculate the energy (per unit time) that is converted to heat in the liquid inside the sphere. This energy W is equal to the mechanical work done on the liquid. If X_n, Y_n, Z_n denote the components of the pressure exerted on the surface of the sphere of radius R, we have

$$W = \int (X_n u + Y_n v + Z_n w)\, ds,$$

where the integral is to be extended over the surface of the sphere of radius R. We have here

$$X_n = -\left(X\xi \frac{\xi}{\rho} + X\eta \frac{\eta}{\rho} + X\zeta \frac{\zeta}{\rho} \right),\ [9]$$

$$Y_n = -\left(Y\xi \frac{\xi}{\rho} + Y\eta \frac{\eta}{\rho} + Y\zeta \frac{\zeta}{\rho} \right),$$

$$Z_n = -\left(Z\xi \frac{\xi}{\rho} + Z\eta \frac{\eta}{\rho} + Z\zeta \frac{\zeta}{\rho} \right),$$

where

$$X_\xi = p - 2k\frac{\delta u}{\delta \xi}, \qquad Y_\zeta = Z_\eta = -k\left(\frac{\delta v}{\delta \zeta} + \frac{\delta w}{\delta \eta}\right),$$

$$Y_\eta = p - 2k\frac{\delta v}{\delta \eta}, \qquad Z_\xi = X_\zeta = -k\left(\frac{\delta w}{\delta \xi} + \frac{\delta u}{\delta \zeta}\right),$$

$$Z_\zeta = p - 2k\frac{\delta w}{\delta \zeta}, \qquad X_\eta = Y_\xi = -k\left(\frac{\delta u}{\delta \eta} + \frac{\delta v}{\delta \xi}\right).$$

The expressions for u, v, w become simpler if we take into account that for $\rho = R$ the terms with the factor P^5/ρ^5 vanish in comparison to those with the factor P^3/ρ^3. We have to set

$$
\begin{cases}
u = A\xi - \tfrac{5}{2}P^3\dfrac{\xi(A\xi^2 + B\eta^2 + C\zeta^2)}{\rho^5}, \\[2mm]
v = B\eta - \tfrac{5}{2}P^3\dfrac{\eta(A\xi^2 + B\eta^2 + C\zeta^2)}{\rho^5}, \qquad (6a)^{[10]}\\[2mm]
w = C\zeta - \tfrac{5}{2}P^3\dfrac{\zeta(A\xi^2 + B\eta^2 + C\zeta^2)}{\rho^5}.
\end{cases}
$$

For p we obtain from the first of equations (5), by similar neglect of terms,

$$p = -5kP^3\frac{A\xi^2 + B\eta^2 + C\zeta^2}{\rho^5} + \text{const.}^{[11]}$$

Now we obtain

$$X_\xi = -2kA + 10kP^3\frac{A\xi^2}{\rho^5} - 25kP^3\frac{\xi^2(A\xi^2 + B\eta^2 + C\zeta^2)}{\rho^7}^{[12]}$$

$$X_\eta = +10kP^3\frac{A\xi\eta}{\rho^5} - 25kP^3\frac{\eta^2(A\xi^2 + B\eta^2 + C\zeta^2)}{\rho^7}^{[13]}$$

$$X_\zeta = +10kP^3\frac{A\xi\zeta}{\rho^5} + 25kP^3\frac{\zeta^2(A\xi^2 + B\eta^2 + C\zeta^2)}{\rho^7},$$

and from this,

$$X_n = 2Ak\frac{\xi}{\rho} - 10AkP^3\frac{\xi}{\rho^4} + 25kP^3\frac{\xi(A\xi^2 + B\eta^2 + C\zeta^2)}{\rho^6}. \quad [14]$$

With the help of the expressions for Y_n and Z_n derived by cyclic permutation, and ignoring all terms that contain the ratio P/ρ in higher than the third power, we get[15]

$$X_n u + Y_n v + Z_n w + \frac{2k}{\rho}(A^2\xi^2 + B^2\eta^2 + C^2\zeta^2)$$
$$-10k\frac{P^3}{\rho^4}(A^2\xi^2 + . + .) + 20k\frac{P^3}{\rho^6}(A\xi^2 + . + .)^2.$$

If we integrate over the sphere and take into account that

$$\int ds = 4R^2\pi,$$

$$\int \xi^2\,ds = \int \eta^2\,ds = \int \zeta^2\,ds = \tfrac{4}{3}\pi R^4,$$

$$\int \xi^4\,ds = \int \eta^4\,ds = \int \zeta^4\,ds = \tfrac{4}{5}\pi R^6,$$

$$\int \eta^2\zeta^2\,ds = \int \zeta^2\xi^2\,ds = \int \xi^2\eta^2\,ds = \tfrac{4}{15}\pi R^6, [16]$$

$$\int (A\xi^2 + B\eta^2 + C\zeta^2)^2\,ds = \tfrac{4}{15}\pi R^6(A^2 + B^2 + C^2), [17]$$

we get[18]

$$W = \tfrac{8}{3}\pi R^3 k\delta^2 - \tfrac{8}{3}\pi P^3 k\delta^2 = 2\delta^2 k(V - \Phi), \quad (7)$$

where we set

$$\delta = A^2 + B^2 + C^2, [19]$$
$$\tfrac{4}{3}\pi R^3 = V$$

and

$$\tfrac{4}{3}\pi P^3 = \Phi.$$

If the suspended sphere were not present ($\Phi = 0$), we would obtain

$$W_0 = 2\delta^2 kV \qquad (7a)$$

for the energy dissipated in volume V. Thus, the presence of the sphere decreases the energy dissipated by $2\delta^2 k\Phi$. It is noteworthy that the effect of the suspended sphere on the quantity of energy dissipated is exactly the same as it would be if the presence of the sphere did not affect the motion of the liquid around it at all.[20]

2. Calculation of the Coefficient of Viscosity of a Liquid in Which Very Many Irregularly Distributed Small Spheres Are Suspended

In the previous section we considered the case where, in a region G of the order of magnitude defined earlier, a sphere is suspended that is very small compared with the region, and we investigated how this sphere affects the motion of the liquid. We are now going to assume that region G contains innumerably many randomly distributed spheres of equal radius, and that this radius is so small that the combined volume of all of the spheres is very small compared to the region G. Let the number of spheres per unit volume be n, where, up to negligibly small terms, n is constant throughout the liquid.

Again, we begin with the motion of a homogeneous liquid without any suspended spheres and consider again the most general dilational motion. If no spheres are present, an appropriate choice of the coordinate system will permit us to represent the velocity components u_0, v_0, w_0 at an arbitrary point x, y, z of G by the equations

$$u_0 = Ax,$$
$$v_0 = By,$$
$$w_0 = Cz,$$

where

$$A + B + C = 0.$$

A sphere suspended at point x_v, y_v, z_v will affect this motion in a way that is evident from equation (6).[21] Since we are choosing the average distance between neighboring spheres to be large compared to their radius, and consequently the additional velocity components arising from all the suspended spheres are very small compared to u_0, v_0, w_0, we obtain for the velocity components u, v, w in the liquid, after taking into account the suspended spheres and neglecting terms of higher orders,

$$\left\{ \begin{aligned} u &= Ax - \sum \left\{ \begin{aligned} &\frac{5}{2}\frac{P^3}{\rho_\nu^2}\frac{\xi_\nu(A\xi_\nu^2 + B\eta_\nu^2 + C\zeta_\nu^2)}{\rho_\nu^3} \\ &-\frac{5}{2}\frac{P^5}{\rho_\nu^4}\frac{\xi_\nu(A\xi_\nu^2 + B\eta_\nu^2 + C\zeta_\nu^2)}{\rho_\nu^3} + \frac{P^5}{\rho_\nu^4}\frac{A\xi_\nu}{\rho_\nu} \end{aligned} \right\}, \\ v &= By - \sum \left\{ \begin{aligned} &\frac{5}{2}\frac{P^3}{\rho_\nu^2}\frac{\eta_\nu(A\xi_\nu^2 + B\eta_\nu^2 + C\zeta_\nu^2)}{\rho_\nu^3} \\ &-\frac{5}{2}\frac{P^5}{\rho_\nu^4}\frac{\eta_\nu(A\xi_\nu^2 + B\eta_\nu^2 + C\zeta_\nu^2)}{\rho_\nu^3} + \frac{P^5}{\rho_\nu^4}\frac{B\eta_\nu}{\rho_\nu} \end{aligned} \right\}, \\ w &= Cz - \sum \left\{ \begin{aligned} &\frac{5}{2}\frac{P^3}{\rho_\nu^2}\frac{\zeta_\nu(A\xi_\nu^2 + B\eta_\nu^2 + C\zeta_\nu^2)}{\rho_\nu^3} \\ &-\frac{5}{2}\frac{P^5}{\rho_\nu^4}\frac{\zeta_\nu(A\xi_\nu^2 + B\eta_\nu^2 + C\zeta_\nu^2)}{\rho_\nu^3} + \frac{P^5}{\rho_\nu^4}\frac{C\zeta_\nu}{\rho_\nu} \end{aligned} \right\}, \end{aligned} \right. \tag{8}$$

where the sum is to be extended over all spheres in the region G and where we have set

$$\xi_\nu = x - x_\nu,$$
$$\eta_\nu = y - y_\nu, \qquad \rho_\nu = \sqrt{\xi_\nu^2 + \eta_\nu^2 + \zeta_\nu^2}.$$
$$\zeta_\nu = z - z_\nu,$$

x_ν, y_ν, z_ν are the coordinates of the centers of the spheres. Furthermore, from equations (7) and (7a) we conclude that, up to infinitesimally small quantities of higher order, the presence of each sphere results in a decrease of heat production by $2\delta^2 k\Phi$ per unit time[22] and that the energy converted to heat in region G has the value

$$W = 2\delta^2 k - 2n\delta^2 k\Phi$$

per unit volume, or

$$W = 2\delta^2 k(1 - \varphi), \tag{7b}$$

where φ denotes the fraction of the volume that is occupied by the spheres.

Equation (7b) gives the impression that the coefficient of viscosity of the inhomogeneous mixture of liquid and suspended spheres (in the following called "mixture" for short) is smaller than the coefficient of viscosity k of the liquid.[23] However, this is not so, since A, B, C are not the values of the principal dilations of the liquid flow represented by equations (8); we will call the principal dilations of the mixture A^*, B^*, C^*. For reasons of symmetry, it follows that the directions of the principal dilations of the mixture are parallel to the directions of the principal dilations A, B, C, i.e., to the coordinate axes. If we write equations (8) in the form

$$u = Ax + \sum u_\nu,$$
$$v = By + \sum v_\nu,$$
$$w = Cz + \sum w_\nu,$$

we get

$$A^* = \left(\frac{\delta u}{\delta x}\right)_{x=0} = A + \sum \left(\frac{\delta u_\nu}{\delta x}\right)_{x=0} = A - \sum \left(\frac{\delta u_\nu}{\delta x_\nu}\right)_{x=0}.$$

If we exclude the immediate surroundings of the individual spheres, we can omit the second and third terms in the expressions for u, v, w and thus obtain for $x = y = z = 0$:

$$\begin{cases} u_\nu = -\frac{5}{2}\frac{P^3}{r_\nu^2}\frac{x_\nu(Ax_\nu^2 + By_\nu^2 + Cz_\nu^2)}{r_\nu^3}, \\[2mm] v_\nu = -\frac{5}{2}\frac{P^3}{r_\nu^2}\frac{y_\nu(Ax_\nu^2 + By_\nu^2 + Cz_\nu^2)}{r_\nu^3}, \\[2mm] w_\nu = -\frac{5}{2}\frac{P^3}{r_\nu^2}\frac{x(Ax_\nu^2 + By_\nu^2 + Cz_\nu^2)}{r_\nu^3}, \end{cases} \quad (9)[24]$$

where we have set

$$r_\nu = \sqrt{x_\nu^1 + y_\nu^2 + z_\nu^2} > 0.$$

We extend the summation over the volume of a sphere K of very large radius R whose center lies at the coordinate origin. Further, if we consider the *irregularly* distributed spheres as being *uniformly* distributed and replace the sum with an integral, we obtain[25]

$$A^* = A - n\int_K \frac{\delta u_\nu}{\delta x_\nu}\, dx_\nu\, dy_\nu\, dz_\nu,$$
$$= A - n\int \frac{u_\nu x_\nu}{r_\nu}\, ds,$$

where the last integral extends over the surface of the sphere K. Taking into account (9), we find that

$$A^* = A - \frac{5}{2}\frac{P^3}{R^6}n \int x_0^2(Ax_0^2 + By_0^2 + Cz_0^2)\,ds,$$
$$= A - n\left(\tfrac{4}{3}P^3\pi\right)A = A(1 - \varphi).$$

Analogously,

$$B^* = B(1 - \varphi),$$
$$C^* = C(1 - \varphi).$$

If we set

$$\delta^{*2} = A^{*2} + B^{*2} + C^{*2},^{[26]}$$

then, neglecting infinitesimally small terms of higher order,

$$\delta^{*2} = \delta^2(1 - 2\varphi).$$

For the heat developed per unit time and volume, we found[27]

$$W^* = 2\delta^2 k(1 - \varphi).$$

If k^* denotes the coefficient of viscosity of the mixture, we have

$$W^* = 2\delta^{*2}k^*.$$

The last three equations yield, neglecting infinitesimal quantities of higher order,

$$k^* = k(1 + \varphi).^{[28]}$$

Thus we obtain the following result:

If very small rigid spheres are suspended in a liquid, the coefficient of viscosity increases by a fraction that is equal to the total volume of the spheres suspended in a unit volume, provided that this total volume is very small.[29]

3. On the Volume of a Dissolved Substance
Whose Molecular Volume Is Large
Compared to That of the Solvent

Consider a dilute solution of a substance that does not dissociate in the solution. Let a molecule of the dissolved substance be large compared to a molecule of the solvent and be considered as a rigid sphere of radius P. We can then apply the result obtained in section 2. If k^* denotes the coefficient of viscosity of the solution and k that of the pure solvent, we have

$$\frac{k^*}{k} = 1 + \varphi,$$

where φ is the total volume of the molecules per unit volume of the solution.[30]

We wish to calculate φ for a 1% aqueous solution of sugar. According to Burkhard's observations (Landolt and Börnstein's *Tables*), $k^*/k = 1.0245$ (at 20°C) for a 1% aqueous sugar solution, hence $\varphi = 0.0245$ for (almost exactly) 0.01 g of sugar. Thus, one gram of sugar dissolved in water has the same effect on the coefficient of viscosity as do small suspended rigid spheres of a total volume of 2.45 cm³.[31] This consideration neglects the effect exerted on the viscosity of the solvent by the osmotic pressure resulting from the dissolved sugar.

Let us remember that 1 g of solid sugar has a volume of 0.61 cm³. This same volume is also found for the specific volume s of sugar in solution if one considers the sugar solution as a *mixture* of water and sugar in dissolved form. I.e., the density of a 1% aqueous sugar solution (relative to water of the same temperature) at 17.5° is 1.00388. Hence we have (neglecting the difference between the density of water at 4° and at 17.5°)

$$\frac{1}{1.00388} = 0.99 + 0.01\,s,$$

and thus

$$s = 0.61.$$

Thus, while the sugar solution behaves like a mixture of water and solid sugar with respect to its density, the effect on viscosity is four times larger than what would result from the suspension of the same amount of sugar.[32] It seems to me that, from the point of view of molecular theory, this result can only be interpreted by assuming that a sugar molecule in solution impedes the mobility of the water in its immediate vicinity, so that an amount of water whose volume is about three times larger than the volume of the sugar molecule is attached to the sugar molecule.[33]

Hence we may say that a dissolved molecule of sugar (i.e., the molecule together with the water attached to it) behaves hydrodynamically like a sphere with a volume of $2.45 \cdot 342/N$ cm^3, where 342 is the molecular weight of sugar and N is the number of actual molecules in one gram-molecule.[34]

4. On the Diffusion of an Undissociated Substance in a Liquid Solution

Let us consider a solution of the kind discussed in section 3. If a force K acts upon a molecule, which we assume to be a sphere with radius P, the molecule will move with a velocity ω, which is determined by P and the coefficient of viscosity

k of the solvent. Indeed, the following equation holds:[3]

$$\omega = \frac{K}{6\pi kP}. \tag{1}$$

We use this relation to calculate the coefficient of diffusion of an undissociated solution. If p is the osmotic pressure of the dissolved substance, the only motion-producing force in such a dilute solution, then the force acting on the dissolved substance per unit volume of solution in the direction of the X-axis equals $-\delta p/\delta x$. If there are ρ grams per unit volume, and m is the molecular weight of the dissolved substance and N the number of actual molecules in one gram-molecule, then $(\rho/m) \cdot N$ is the number of (actual) molecules per unit volume, and the force exerted on a molecule by virtue of the concentration gradient is

$$K = -\frac{m}{\rho N} \frac{\delta p}{\delta x}. \tag{2}$$

If the solution is sufficiently dilute, the osmotic pressure is given by the equation:

$$p = \frac{R}{m} \rho T, \tag{3}$$

where T is the absolute temperature and $R = 8.31 \cdot 10^7$. From equations (1), (2), and (3) we obtain for the migration velocity of the dissolved substance

$$\omega = -\frac{RT}{6\pi k} \frac{1}{NP} \frac{1}{\rho} \frac{\delta \rho}{\delta x}. \tag{4}$$

---

[3] G. Kirchhoff, *Vorlesungen über Mechanik*, 26. Vorl. (*Lectures on Mechanics*, Lecture 26), equation (22).

Finally, the amount of the substance passing per unit time through a unit cross section in the direction of the X-axis is

$$\omega\rho = -\frac{RT}{6\pi k} \cdot \frac{1}{NP}\frac{\delta\rho}{\delta x}.$$

Hence, we obtain for the coefficient of diffusion D

$$D = \frac{RT}{6nk} \cdot \frac{1}{NP}. \quad [35]$$

Thus, from the coefficients of diffusion and viscosity of the solvent we can calculate the product of the number N of actual molecules in one gram-molecule and the hydrodynamically effective molecular radius P.

In this derivation the osmotic pressure has been treated as a force acting on the individual molecules, which obviously does not agree with the viewpoint of the kinetic molecular theory; since in our case—according to the latter—the osmotic pressure must be conceived as only an apparent force. However, this difficulty disappears when one considers that the (apparent) osmotic forces that correspond to the concentration gradients in the solution may be kept in (dynamic) equilibrium by means of numerically equal forces acting on the individual molecules in the opposite direction, which can easily be seen by thermodynamic methods.

The osmotic force acting on a unit mass $-\frac{1}{\rho}\frac{\delta p}{\delta x}$ can be counterbalanced by the force $-P_x$ (exerted on the individual dissolved molecules) if

$$-\frac{1}{\rho}\frac{\delta p}{\delta x} - P_x = 0.$$

Thus, if one imagines that (per unit mass) the dissolved substance is acted upon by two sets of forces P_x and $-P_x$ that mutually cancel out each other, then $-P_x$ counterbalances the osmotic pressure, leaving only the force P_x, which is numerically equal to the osmotic pressure, as the cause of motion. The difficulty mentioned above has thus been eliminated.[4]

5. Determination of Molecular Dimensions with the Help of the Obtained Relations

We found in section 3 that

$$\frac{k^*}{k} = 1 + \varphi = 1 + n \cdot \tfrac{4}{3}\pi P^3, \text{[36]}$$

where n is the number of dissolved molecules per unit volume and P is the hydrodynamically effective radius of the molecule. If we take into account that

$$\frac{n}{N} = \frac{\rho}{m},$$

where ρ denotes the mass of the dissolved substance per unit volume and m its molecular weight, we get

$$NP^3 = \frac{3}{4\pi} \frac{m}{\rho} \left(\frac{k^*}{k} - 1 \right). \text{[37]}$$

On the other hand, we found in section 4 that

$$NP = \frac{RT}{6\pi k} \frac{1}{D}.$$

These two equations enable us to calculate separately the quantities P and N, of which N must be independent of the nature of the solvent, the dissolved substance, and the temperature, if our theory agrees with the facts.

[4] A detailed presentation of this line of reasoning can be found in *Ann. d. Phys.* 17 (1905): 549. [See also this volume, paper 2, p. 86.]

We will carry out the calculation for an aqueous solution of sugar. From the data on the viscosity of the sugar solution cited earlier, it follows that at 20°C,

$$NP^3 = 200.^{[38]}$$

According to the experiments of Graham (as calculated by Stefan), the diffusion coefficient of sugar in water is 0.384 at 9.5°C, if the day is chosen as the unit of time. The viscosity of water at 9.5° is 0.0135. We will insert these data in our formula for the diffusion coefficient, even though they have been obtained using 10% solutions, and strict validity of our formula cannot be expected at such high concentrations. We obtain

$$NP = 2.08 \cdot 10^{16}.$$

Neglecting the differences between the values of P at 9.5° and 20°, the values found for NP^3 and NP yield

$$P = 9.9 \cdot 10^{-8} \text{ cm},$$
$$N = 2.1 \cdot 10^{23}.$$

The value found for N shows satisfactory agreement, in order of magnitude, with values found for this quantity by other methods.[39]

(Bern, 30 April 1905)

EDITORIAL NOTES

[1]A factor k is missing on the right-hand side of the last equation in this line; this error is corrected in Albert Einstein, "Eine neue Bestimmung der Moleküldimensionen," *Ann. d. Phys.* 19 (1906), pp. 289–305, cited hereafter as *Einstein 1906*. Note that $\frac{\delta}{\delta}$ denotes partial differentiation (modern $\frac{\partial}{\partial}$).

[2]The denominator on the right-hand side should be $\delta\xi^2$; this error is corrected in ibid.

[3]The denominator of the first term on the right-hand side should be $\delta\xi^2$; this error is corrected in ibid. A reprint of this article in the Einstein Archive shows marginalia and interlineations in Einstein's hand, the first of which refer to this and the following equation. The term "$+g\,\frac{1}{\rho}$" was added to the right-hand side of the equations for V and then canceled. These marginalia and interlineations are presumably part of Einstein's unsuccessful attempt to find a calculational error; see note 13 below.

[4]The equation for u' should be, as corrected in ibid., $u' = -2c\frac{\delta\frac{1}{\rho}}{\delta\xi}$. In the reprint mentioned in note 3, the first derivative with respect to ξ was changed to a second derivative and then changed back to a first derivative. At the bottom of the page, the following equations are written:

$$b = -1/12\ P^5 a$$
$$c = -5/12\ P^3 a$$
$$g = 2/3\ P^3 a.$$

[5]The numerator of the last term in the curly parentheses should be "$\delta^2(1/\rho)$," as corrected in ibid.

[6]$\frac{\delta n}{\delta\xi}$ should be $\frac{\delta p}{\delta\xi}$, as corrected in Einstein, *Untersuchungen über die Theorie der 'Brownschen Bewegung'* (ed. Reinhold Fürth. Ostwald's Klassiker der exakten Wissenschaften, no. 199. Leipzig: Akademische Verlagsgesellschaft, 1922); cited hereafter as *Einstein 1922*.

[7]The factor preceding the first parenthesis should be, as corrected in *Einstein 1906,*

$$-5/2\frac{P^3}{\rho^5}.$$

[8]The equations should be $u = U$, $v = V$, $w = W$.

[9]$X\xi$, $X\eta$, $X\zeta$ should be X_ξ, etc., as corrected in *Einstein 1906*.

[10]In Einstein's reprint (see note 3), the term $+\frac{5}{6}\ P^3\ \frac{A\xi}{\rho^3}$ is added to the right-hand side of the first equation. After the last terms of the second and third equations, series of dots are added. These interlineations are presumably related to the marginal calculations indicated in note 3.

[11]In Einstein's reprint (see note 3), the term $+5\,k\,P^3\frac{1}{\rho^3}$ is added to the right-hand side of this equation. This interlineation is presumably related to the marginal calculations referred to in note 3.

[12]In Einstein's reprint (see note 3), the term $-\frac{5}{3} k P^3 A\left(\frac{1}{\rho^3} - 9\frac{\xi}{\rho^5}\right)$ is added to the right-hand side of this equation. This addition is presumably related to the marginal calculations referred to in note 3.

[13]This equation and the subsequent one are incorrect. Apart from minor errors, they contain a calculational error bearing on the numerical factors. In *Einstein 1906*, +25 in front of the last term in the equation for X_ζ is changed to −25. In Einstein's reprint (see note 3), the factor ζ^2 in the last term on the right-hand side of this equation is corrected to $\xi\zeta$, and the factor η^2 in front of the parenthesis in the last term on the right-hand side of the equation for X_η is corrected to $\xi\eta$. The calculational error that is also contained in these equations, and some of its consequences, are corrected in "Berichtigung zu meiner Arbeit: 'Eine Neue Bestimmung der Moleküldimensionen,'" *Collected Papers*, vol. 3, doc. 14, pp. 416–417. The corrections are integrated into the text of the reprint of this paper in *Einstein 1922*. The correct equations are:

$$X_\eta = +5kP^3 \frac{(A + B)\xi\eta}{\rho^5}$$
$$-25kP^3 \frac{\xi\eta(A\xi^2 + B\eta^2 + C\zeta^2)}{\rho^7}$$
$$X_\zeta = +5kP^3 \frac{(A + C)\xi\zeta}{\rho^5}$$
$$-25kP^3 \frac{\xi\zeta(A\xi^2 + B\eta^2 + C\zeta^2)}{\rho^7}.$$

[14]−10 should be replaced by −5 and 25 by 20 (see previous note).

[15]The third + sign should be replaced by = as corrected in *Einstein 1922*. −10 should be replaced by −5, and 20 by 15 (see note 13).

[16]In Einstein's reprint (see note 3), the factor 4/15 was changed to 8/15 and then changed back to 4/15.

[17]4/15 should be replaced by 8/15 as corrected in Einstein's reprint (see note 3).

[18]This equation should be (see note 13):

$$W = 8/3\pi R^3 k\delta^2 + 4/3\pi P^3 k\delta^2$$
$$= 2\delta^2 k(V + \Phi/2).$$

[19]δ should be δ^2. This correction is made in Einstein's reprint (see note 3).

[20]It follows from the correction to eq. (7) that the dissipated energy is actually increased by half this amount. The statement in the text is only partially corrected in *Einstein 1922*; the amount is correctly given but still

described as a diminution. The final sentence of this paragraph, which no longer applies to the corrected calculation, is omitted from *Einstein 1922*.

[21]The point should be denoted by x_ν, y_ν, z_ν, as corrected in *Einstein 1906*.

[22]The heat production per unit time is actually increased by $\delta^2 k\Phi$. The correct equations are thus (see note 13): $W = 2\delta^2 k + n\delta^2 k\Phi$, and $W = 2\delta^2 k(1 + \Phi/2)$.

[23]The following two sentences are revised in *Einstein 1922*: "In order to calculate from equation (7b) the coefficient of friction of an inhomogeneous mixture of fluid and suspended spheres (in the following called 'mixture' for short) that we are examining, we must further take into consideration that A, B, C are not values of the principal dilations of the motion of fluid represented in equation (8); we want to designate the principal dilations of the mixture as A^*, B^*, C^*."

[24]In this and the following two equations, the sign after $=$ should be $+$; the third equation should have z_ν instead of x in the numerator; the latter correction is made in *Einstein 1906*.

[25]The factor in front of the second term in the first equation is 5/2 (see ibid.). In deriving the second equation, Einstein used the equations in the middle of p. 54 and the fact that $A + B + C = 0$.

[26]In Einstein's reprint (see note 3), $= A^2 + B^2 + \delta^2(1 - 2\varphi)$ is added to the right-hand side of this equation and then crossed out.

[27]The correct equation is (see note 13): $W^* = 2\delta^2 k(1 + \varphi/2)$.

[28]The correct equation is (see note 13): $k^* = k(1 + 2.5\varphi)$.

[29]The fraction is actually 2.5 times the total volume of the suspended spheres (see note 13).

[30]The correct equation is (see note 13): $k^*/k = 1 + 2.5\varphi$.

[31]The correct value is 0.98 cm^3 (see note 13). The following sentence is omitted in *Einstein 1906*.

[32]The viscosity is actually one and one-half times greater (see note 13).

.[33]The quantity of water bound to a sugar molecule has a volume that is actually one-half that of the sugar molecule (see note 13). The existence of molecular aggregates in combination with water was debated at that time.

[34]The volume of the sphere is actually $0.98 \cdot 342/N$ cm^3 (see note 13).

[35]The first denominator should be $6\pi k$, as corrected in *Einstein 1906*. This equation was obtained independently by William Sutherland in 1905 by a similar argument. The idea to use this formula for a determination of molecular dimensions may have occurred to Einstein as early as 1903.

[36]The correct equation is (see note 13): $k^*/k = 1 + 2.5\varphi = 1 + 2.5n\, 4/3\, \pi P^3$.

[37]The correct equation has an additional factor 2/5 on the right-hand side (see note 13).

[38]For the experimental data, see p. 60. The correct value is 80 (see note 13).

[39]The values obtained by using the correct equations (see *Einstein 1922*) are $P = 6.2\ 10^{-8}$ cm; and $N = 3.3\ 10^{23}$ (per mole).

Einstein as a student at the ETH, or shortly thereafter.
(Courtesy of Hebrew University of Jerusalem)

Einstein's study of Brownian motion constitutes one of the high points in the long tradition of research on the kinetic theory of heat and of his own contributions to this field. Some of the consequences of his work were of great significance for the development of physics in the twentieth century. Einstein's derivation of the laws governing Brownian motion, and their subsequent experimental verification by Perrin and others, contributed significantly to the acknowledgment of the physical reality of atoms by the then still-numerous skeptics. His papers on Brownian motion helped to establish the study of fluctuation phenomena as a new branch of physics. The methods he created in the course of his research prepared the way for statistical thermodynamics, later developed by Szilard and others, and for a general theory of stochastic processes.

Since at least the middle of the nineteenth century, a growing number of physicists and chemists had accepted the atomic hypothesis. The assumption that matter consists of atoms and molecules suggested a number of relations between phenomena, both physical and chemical, that are unexpected from a purely macroscopic point of view. Various methods for the determination of molecular dimensions gave values that were often in surprisingly good agreement. The physical reality of atoms was not, however, universally accepted by the end of the century. There were still some fervent opponents of the atomic hypothesis, such as Wilhelm Ostwald and Georg Helm, who called themselves "energeticists" to indicate that they regarded the concept of energy as the most fundamental ontological concept of science. Others, such as Ernst Mach, while adopting a hostile position with regard to the existence of entities not directly accessible to sense experience—in particular, atoms—admitted that atomism may have a heuristic or didactic utility. It was not uncommon, even among scientists who made explicit use of atomic assumptions in their work, to regard atomism as a mere working hypothesis.

Although at the turn of the century the atomic hypothesis was proving its heuristic value in such new areas of research as the electron theory of metals and stereochemistry, some physicists had come to regard the theory of heat as an area in which the atomic hypothesis was no longer fruitful. Einstein probably became aware of the controversy over the molecular theory of heat during his student years, when he read works by Mach, Ostwald, and Boltzmann. In 1900 Einstein finished reading Boltzmann's *Gastheorie* (1896, 1898), in which Boltzmann, presumably reacting to a dispute with Ostwald and Helm, suggested that he was isolated in his support of the kinetic theory. Although he criticized Boltzmann for a lack of emphasis on the comparison of his theory with observation, Einstein was firmly convinced of the principles of Boltzmann's theory.

In his first published attempts at independent research, Einstein took for granted the atomistic constitution of matter and of electricity. He developed a theory of molecular forces, on the basis of which he established a number of relations among observable phenomena. Einstein's interest soon shifted from the details of molecular forces to the quest for facts, "which would guarantee as much as possible the existence of atoms of definite finite size," as he later characterized this phase of his work.[1]

THE IRREGULAR movement of microscopic particles suspended in a liquid had been noted long before the botanist

Robert Brown published his careful observations in 1828, but Brown was the first to emphasize its ubiquity and to exclude its explanation as a vital phenomenon. Advances in observational technique and in theory served to eliminate a number of unsatisfactory explanations of Brownian motion by the end of the nineteenth century, if not to verify the correct one. Explanations of Brownian motion proposed after the exclusion of vital forces involved capillarity, convection currents, evaporation, interaction with light, and electrical forces. During the 1870s, the kinetic theory of heat was proposed as an explanation by several authors. A powerful argument against this explanation was developed by the cytologist Karl von Nägeli in 1879. He first used the equipartition theorem to calculate the average velocity of the molecules of the liquid, and then used the laws of elastic collision to obtain the velocity of a suspended particle. He concluded that the velocity of such a particle, because of its comparatively large mass, would be vanishingly small. William Ramsay and Louis-Georges Gouy independently tried to defend the molecular explanation of Brownian motion by assuming the existence of collective motions of large numbers of atoms in liquids, an assumption suited to the refutation of arguments such as Nägeli's.

In 1900 an entirely different way of applying the kinetic theory of heat to Brownian motion was investigated by Felix Exner, who assumed an equipartition of energy between the molecules of the liquid and the suspended particles. He calculated the velocity of the molecules on the basis of observations that he interpreted as giving the mean velocities of the suspended particles, obtaining results that were not in agreement with contemporary estimates of molecular velocities. In Exner's work there is no fundamental difference between a solute molecule and a suspended particle. Einstein arrived at a similar conclusion, but instead of emphasizing the equipartition theorem, he took the osmotic pressure and its relation to the theory of diffusion and to the molecular theory of heat as the starting point of his analysis of Brownian motion. He writes in the paper that follows: "According to this theory, a dissolved molecule differs from a suspended body in size, *only* and it is difficult to see why suspended bodies should not produce the same osmotic pressure as an equal number of dissolved molecules" (see pp. 86–87).

On the other hand, Einstein pointed out, according to the "classical theory of thermodynamics," suspended particles—as macroscopic objects—should not exert an osmotic pressure on a semipermeable wall. Before Einstein, no one seems to have recognized that this contrast provides a touchstone for the kinetic theory. His choice of a suspension to study the relations between the thermodynamic and atomic theories of heat amounted to a radical reversal of perspective. Usually the legitimacy of microscopic explanations of thermodynamic results was at issue. In this case, however, the question centered on the applicability of a thermodynamic concept—osmotic pressure—to the suspended particles.

In the course of studying colloidal solutions, the commonly made distinction between suspensions and solutions in nineteenth-century chemistry had lost its absolute character. The absence of any fundamental difference between solutions and suspensions was made strikingly clear in 1902, when observations performed with the newly invented ultramicroscope made it possible to resolve many colloidal solutions into their constituents. The ultramicroscope not only demonstrated the physical reality of colloidal particles, but showed that irregular motion is one of their outstanding characteristics.

Although the ultramicroscope brought closer what Jean Perrin called the "distant reality" of molecules, one of their fundamental properties, their velocities, remained inaccessible to measurement. The inconsistencies that result from presumed velocity measurements, such as Exner's, had hinted at this problem; but it was explicitly discussed for the first time in the theoretical studies of Brownian motion that Einstein and Smoluchowski independently published between 1905 and 1907. Both introduced the mean-square displacement of the suspended particles as the primary observable quantity in Brownian motion. Einstein argued that dissipative forces change the direction and magnitude of the velocity of a suspended particle on such a short timescale that it cannot be measured. This argument demonstrates the fundamental role of dissipation in Einstein's analysis of Brownian motion.

In summary, the study of previous explanations of Brownian motion shows that three elements of Einstein's approach are characteristic of his decisive progress: (1) he based his analysis on the osmotic pressure rather than on the equipartition theorem; (2) he identified the mean-square displacements of suspended particles rather than their velocities as suitable observable quantities; and (3) he simultaneously applied the molecular theory of heat and the macroscopic theory of dissipation to the same phenomenon, rather than restricting each of these conceptual tools to a single scale, molecular or macroscopic.

IN THE PAPER on Brownian motion (paper 2), Einstein proved "that, on the assumption of the molecular theory of heat, bodies of the order of magnitude of 1/1000 mm suspended in liquids must already carry out an observable random movement, which is generated by thermal motion," as he wrote in a letter to Conrad Habicht in the late spring of 1905. Einstein wrote this paper "without knowing that observations concerning Brownian motion were already long familiar."[2] He did not mention Brownian motion in the title, although he conjectured that the motion he predicted might be identical to Brownian motion. Boltzmann's *Gastheorie*, which Einstein carefully studied during his student years, explicitly denies that the thermal motion of molecules in a gas leads to observable motions of suspended bodies. (This denial may be an instance of what Einstein referred to as Boltzmann's attaching too little importance to a comparison of theory with observation.) Sometime between 1902 and 1905, Einstein read Poincaré's *La science et l'hypothèse*, which contains a brief discussion of Gouy's work on Brownian motion, emphasizing Gouy's argument that Brownian motion violates the second law of thermodynamics. Einstein's second paper on Brownian motion,[3] written after Siedentopf drew his attention to Gouy's work, cites Gouy's observations as qualitative confirmation of his results.

Paper 2 opens with the derivation of an expression for the diffusion coefficient in terms of the radius of the suspended particles and the temperature and viscosity of the liquid, an expression already obtained in Einstein's doctoral dissertation. Unlike the previous derivation, however, the new one makes use of the methods of statistical physics that Einstein developed. The new approach is different in two respects:

1. In his dissertation, which deals with solutions rather than suspensions, Einstein simply assumed the validity of van't

Hoff's law for the osmotic pressure. He now gave a derivation of this law from an expression for the free energy of the suspension that follows from statistical mechanics.

2. Rather than simply considering the equilibrium of forces acting on a single molecule, Einstein derived the equilibrium between the osmotic pressure and a friction force obeying Stokes's law by a thermodynamical argument.

The ensuing derivation of the diffusion equation is based on the introduction of a probability distribution for displacements. The introduction of such a distribution is presumably related to Einstein's previous use of probability distributions. Einstein assumed the existence of a time interval, short with respect to the observation time, yet sufficiently long that the motions of a suspended particle in two successive time intervals can be treated as independent of each other. The displacement of the suspended particles can then be described by a probability distribution that determines the number of particles displaced by a certain distance in each time interval. Einstein derived the diffusion equation from an analysis of the time-dependence of the particle distribution, calculated from the probability distribution for displacements. This derivation is based on his crucial insight into the role of Brownian motion as the microscopic process responsible for diffusion on a macroscopic scale. Compared to such a derivation, one based on an analogy to the treatment of diffusion in the kinetic theory of gases may have appeared more problematic to Einstein because of the lack of a fully developed kinetic theory of liquids.

The solution of the resulting diffusion equation, combined with his expression for the diffusion coefficient, yields an expression for the mean-square displacement, λ_x, as a function of time, an expression that Einstein suggested could be used experimentally to determine Avogadro's number N:

$$\lambda_x = \sqrt{t}\sqrt{\frac{RT}{N}\frac{1}{3\pi kP}}, \tag{1}$$

where t is the time, R the gas constant, T the temperature, k the viscosity, and P the radius of the suspended particles.

Through his earlier work, Einstein was familiar with the theory of diffusion in both gases and liquids, as well as with other techniques needed for his analysis of Brownian motion. In 1902 he suggested the replacement of semipermeable walls in thermodynamic arguments by external conservative forces, a method he stated to be particularly useful for treating arbitrary mixtures. During 1903 Einstein discussed the notions of semipermeable membrane and osmotic pressure in his correspondence with Michele Besso, showing interest in Sutherland's hypothesis on the mechanism of semipermeable membranes. In his papers on statistical physics, Einstein generalized the idea of external conservative forces and noted the significant role of fluctuations in statistical physics. In 1904 he derived an expression for mean square deviations from the average value of the energy of a system.

SEVERAL newly perfected techniques for the experimental investigation of Brownian motion, notably the ultramicroscope and new methods for preparing colloidal solutions, were available by the time Einstein published his first articles on the subject. As one of the first applications of the ultramicroscope, Siedentopf and Zsigmondy observed Brownian motion in colloidal solutions but did not perform precise measurements. *Les ultramicroscopes et les objets ultramicroscopiques*, a book on the ultramicroscope and its applications by Aimé Cotton and Henri Mouton, published in 1906, helped stimulate interest in Brownian motion and brought Einstein's theory to the attention of researchers in the field. Using an ultramicroscope and a sophisticated observational technique, The Svedberg carried out careful measurements of Brownian motion with the aim of testing the interpretation of Brownian motion that maintains it is caused by the thermal motions of molecules. Svedberg reported on his attempts to test Einstein's theory in 1906 and corresponded with Einstein on the subject of Brownian motion, sending him one of his papers.

Svedberg followed Zsigmondy in assuming two types of motions for colloidal particles, a translational motion and a "proper [Brownian] motion." Svedberg restricted his attention to the latter and attempted to facilitate its measurement by superimposing a translational motion. He described the resulting trajectories as "sinusoid-like," but cautioned against concluding that the motion had an oscillatory character.[4] In the analysis of his results, however, Svedberg introduced a terminology that was adapted to the description of a simple oscillatory motion, relating the observed amplitudes to Einstein's root-mean-square displacement. Earlier, he had tried to estimate molecular velocities on the basis of the observed velocities of colloidal particles. In another article, written mainly to correct the basic misunderstandings in Svedberg's work, Einstein showed that the velocities of ultramicroscopic particles, as calculated from the equipartition theorem, cannot be directly observed.[5] On 11 November 1909, Einstein wrote to Perrin: "The errors in Svedberg's method of observation and also in his theoretical treatment became clear to me at once. I wrote a minor correction at the time, which only addressed the worst ones, as I couldn't bring myself to detract from Mr. S.'s great pleasure in his work."

In addition to the misconceptions underlying Svedberg's experimental work, his numerical results slightly disagreed with Einstein's predictions. Other early experimental work on Einstein's and Smoluchowski's theories, such as Felix Ehrenhaft's observations of displacements of aerosol particles, Victor Henri's cinematographical measurements of displacements of suspended particles, or Max Seddig's study of the temperature dependence of Brownian motion, provided qualitative confirmation of the theory; but the work of Henri and Seddig also failed to yield quantitative agreement. As a consequence, the kinetic interpretation of Brownian motion was not universally accepted in 1908 as the exclusive explanation of the phenomenon.

Einstein noted that the control of the temperature was the principal difficulty in obtaining satisfactory results from the photographic records of Seddig and Henri. Before Jean Perrin published his thorough experimental investigation of the phenomenon, Einstein was skeptical about the possibility of obtaining precise measurements of Brownian motion. On 30 July 1908, in a letter to Jakob Laub, Einstein commented rather enthusiastically on Seddig's work in spite of its shortcomings: "I have read Seddig's paper. He has done it very well. I cannot quite make head or tail of his descriptions of the results." Einstein wrote to Jean Perrin on 11 November the following year: "I would have considered it impossible to investigate Brownian motion so precisely; it is a stroke of luck for this subject that you have taken it up."

In a series of experiments, the first results of which were published in 1908, Perrin achieved an until then unmatched precision in the confirmation of almost all of Einstein's predictions. Like Einstein, Perrin recognized that the analogy established by van't Hoff between an ideal gas and a solution could be extended to colloidal solutions and suspensions, and that this analogy provides a unique means of obtaining evidence for atomism. In his first experiments on Brownian motion, Perrin tested a formula for the vertical distribution of suspended particles under the influence of gravitation. Although Perrin probably was aware of Einstein's theory through Langevin, he was apparently unaware that Einstein had derived a similar formula. Challenged by criticism, Perrin checked his assumption of the validity of Stokes's formula for the particles used in his experiments. In two further papers published in 1908, Perrin applied his methods to a determination of Avogadro's number.

In the same year, Perrin's doctoral student Chaudesaigues subjected Einstein's displacement formula to experimental tests. Contrary to Henri's results mentioned above, the results are in excellent agreement with theoretical predictions. Perrin continued these successful experiments with the help of other students; to Einstein's surprise, he was able to include rotational Brownian motion in his investigations. On 11 November 1909, Einstein wrote to Perrin: "I would not have considered a measurement of the rotations as feasible. In my eyes it was only a pretty trifle." Perrin's success was based on the ingenious combination of several experimental techniques for preparing emulsions with precisely controllable particle sizes, and for measuring the particles' number and displacements. He summarized his results in various review articles and books that significantly furthered the general acceptance of atomism.

Beginning in 1907, Einstein himself tried to contribute to the experimental study of fluctuation phenomena. His prediction of voltage fluctuations in condensers stimulated him to explore the possibility of measuring small quantities of electricity in order to provide experimental support for "a phenomenon in the field of electricity related to Brownian motion."[6] On 15 July 1907, he wrote to his friends Conrad and Paul Habicht about his discovery of a method for the measurement of small quantities of electrical energy. Soon afterward the Habichts tried to build the device proposed by Einstein. At the end of 1907, Einstein dropped his idea of obtaining a patent for the device, as he wrote on 24 December to Conrad Habicht, "primarily because of the lack of interest by manufacturers." Instead, he published a paper on the basic features of his method, which stimulated further work on the device.[7] While the use of the device for measuring fluctuation phenomena in conductors proved to be difficult, experimental work done by others soon provided evidence for the atomistic constitution of matter and electricity that exceeded Einstein's initial expectations.

EDITORIAL NOTES

[1] Einstein, *Autobiographical Notes*, Paul Arthur Schilpp, trans. and ed. (La Salle, Ill.: Open Court, 1979), pp. 44–45.

[2] Ibid.

[3] *Annalen der Physik* 19 (1906): 371–381, reprinted in *Collected Papers*, vol. 2, doc. 32, pp. 334–344.

[4] Svedberg, *Zeitschrift für Elektrochemie und angewandte physikalische Chemie* 12 (1906): 853–854.

[5] Einstein, in ibid., 13 (1907): 41–42; reprinted in *Collected Papers*, vol. 2, doc. 40, pp. 399–400.

[6] *Annalen der Physik* 22 (1907): 569–572, reprinted in *Collected Papers*, vol. 2, doc. 39, pp. 393–396.

[7] *Physikalische Zeitschrift* 9 (1908): 216–217, reprinted in *Collected Papers*, vol. 2, doc. 48, pp. 490–491.

PAPER 2

✦

On the Motion of Small Particles Suspended in Liquids at Rest Required by the Molecular-Kinetic Theory of Heat

IN THIS PAPER it will be shown that, according to the molecular-kinetic theory of heat, bodies of a microscopically visible size suspended in liquids must, as a result of thermal molecular motions, perform motions of such magnitude that they can be easily observed with a microscope. It is possible that the motions to be discussed here are identical with so-called Brownian molecular motion; however, the data available to me on the latter are so imprecise that I could not form a judgment on the question.

If the motion to be discussed here can actually be observed, together with the laws it is expected to obey, then classical thermodynamics can no longer be viewed as applying to regions that can be distinguished even with a microscope, and an exact determination of actual atomic sizes becomes possible. On the other hand, if the prediction of this motion were to be proved wrong, this fact would provide a far-reaching argument against the molecular-kinetic conception of heat.

1. On the Osmotic Pressure to Be Ascribed to Suspended Particles

Let z gram-molecules of a non-electrolyte be dissolved in a part V^* of the total volume V of a liquid. If the volume V^* is separated from the pure solvent by a wall that is permeable to the solvent but not to the solute, then this wall is subjected to a so-called osmotic pressure, which for sufficiently large values of V^*/z satisfies the equation

$$pV^* = RTz.$$

But if instead of the solute, the partial volume V^* of the liquid contains small suspended bodies that also cannot pass through the solvent-permeable wall, then, according to the classical theory of thermodynamics, we should not expect— at least if we neglect the force of gravity, which does not interest us here—any pressure to be exerted on the wall; for according to the usual interpretation, the "free energy" of the system does not seem to depend on the position of the wall and of the suspended bodies, but only on the total mass and properties of the suspended substance, the liquid, and the wall, as well as on the pressure and temperature. To be sure, the energy and entropy of the interfaces (capillary forces) should also be considered when calculating the free energy; but we can disregard them here because changes in the position of the wall and suspended bodies will not cause changes in the size and state of the contact surfaces.

But a different interpretation arises from the standpoint of the molecular-kinetic theory of heat. According to this theory, a dissolved molecule differs from a suspended body *only* in size, and it is difficult to see why suspended bodies should not produce the same osmotic pressure as an equal number of dissolved molecules. We have to assume that the suspended bodies perform an irregular, albeit very slow, motion in the liquid due to the liquid's molecular motion; if prevented by the wall from leaving the volume V^*, they will exert pressure upon the wall just like molecules in solution. Thus, if n suspended bodies are present in the volume V^*, i.e., $n/V^* = \nu$ in a unit volume, and if neighboring bodies are sufficiently far separated from each other, there will be a corresponding osmotic pressure p of magnitude

$$p = \frac{RT}{V^*} \frac{n}{N} = \frac{RT}{N} \cdot \nu,$$

where N denotes the number of actual molecules per gram-molecule. It will be shown in the next section that the molecular-kinetic theory of heat does indeed lead to this broader interpretation of osmotic pressure.

2. Osmotic Pressure from the Standpoint of the Molecular-Kinetic Theory of Heat[1]

If $p_1 p_2 \ldots p_l$ are state variables of a physical system that completely determine the system's instantaneous state (e.g., the coordinates and velocity components of all atoms of the

---

[1] In this section it is assumed that the reader is familiar with the author's papers on the foundations of thermodynamics (cf. *Ann. d. Phys.* 9 [1902]: 417 and 11 [1903]: 170). Knowledge of these papers and of this section of the present paper is not essential for an understanding of the results in the present paper.

system), and if the complete system of equations for changes of these variables is given in the form

$$\frac{\partial p_\nu}{\partial t} = \varphi_\nu(p_1 \ldots p_l) \quad (\nu = 1, 2, \ldots l),$$

with $\sum \frac{\partial \varphi_\nu}{\partial p_\nu} = 0$, then the entropy of the system is given by the expression

$$S = \frac{\bar{E}}{T} + 2\kappa \ln \int e^{-\frac{E}{2\kappa T}} dp_1 \ldots dp_l.$$

Here T denotes the absolute temperature, $\bar{E}$ the energy of the system, and E the energy as a function of the p_ν. The integral extends over all possible values of p_ν consistent with the conditions of the problem. κ is connected with the constant N mentioned above by the relation $2\kappa N = R$. Hence we get for the free energy F

$$F = -\frac{R}{N} T \ln \int e^{-\frac{EN}{RT}} dp_1 \ldots dp_l = -\frac{RT}{N} \ln B.$$

Let us now imagine a liquid enclosed in volume V; let a part V^* of the volume V contain n solute molecules or suspended bodies, which are retained in the volume V^* by a semipermeable wall; the integration limits of the integral B occurring in the expressions for S and F will be affected accordingly. Let the total volume of the solute molecules or suspended bodies be small compared with V^*. In accordance with the theory mentioned, let this system be completely described by the variables $p_1 \ldots p_l$.

Even if the molecular picture were extended to include *all* details, the calculation of the integral B would be so difficult that an exact calculation of F is hardly conceivable. However, here we only need to know how F depends on the size of the volume V^* in which all the solute molecules or suspended bodies (hereafter called "particles" for brevity) are contained.

Let us call the rectangular coordinates of the center of gravity of the first particle x_1, y_1, z_1, those of the second x_2, y_2, z_2, etc., and those of the last particle x_n, y_n, z_n, and assign to the centers of gravity of the particles the infinitesimally small parallelepiped regions $dx_1 dy_1 dz_1$, $dx_2 dy_2 dz_2 \ldots dx_n dy_n dz_n$, all of which lie in V^*. We want to evaluate the integral occurring in the expression for F, with the restriction that the centers of gravity of the particles shall lie in the regions just assigned to them. In any case, this integral can be put into the form

$$dB = dx_1 dy_1 \ldots dz_n \cdot J,$$

where J is independent of $dx_1 dy_1$, etc., as well as of V^*, i.e., of the position of the semipermeable wall. But J is also independent of the particular choice of the *positions* of the center of gravity regions and of the value of V^*, as will be shown immediately. For if a second system of infinitesimally small regions were assigned to the centers of gravity of the particles and denoted by $dx_1' dy_1' dz_1'$, $dx_2' dy_2' dz_2' \ldots dx_n' dy_n' dz_n'$, and if these regions differed from the originally assigned ones by their position alone, but not by their size, and if, likewise, all of them were contained in V^*, we would similarly have

$$dB' = dx_1' dy_1' \ldots dz_n' \cdot J',$$

where

$$dx_1 dy_1 \ldots dz_n = dx_1' dy_1' \ldots dz_n'.$$

Hence,

$$\frac{dB}{dB'} = \frac{J}{J'}.$$

But from the molecular theory of heat, as presented in the papers cited,[2] it is easily deduced that dB/B and dB'/B are respectively equal to the probabilities that at an arbitrarily chosen moment the centers of gravity of the particles will be found in the regions $(dx_1 \ldots dz_n)$ and $(dx'_1 \ldots dz'_n)$ respectively. If the motions of the individual particles are independent of one another (to a sufficient approximation) and if the liquid is homogeneous and no forces act on the particles, then for regions of the same size the probabilities of the two systems of regions will be equal, so that

$$\frac{dB}{B} = \frac{dB'}{B}.$$

But from this equation and the previous one it follows that

$$J = J'.$$

This proves that J does not depend on either V^* or $x_1, y_1 \ldots z_n$. By integrating, we get

$$B = \int J \, dx_1 \ldots dz_n = JV^{*n},$$

and hence

$$F = -\frac{RT}{N}\{\ln J + n \ln V^*\}$$

and

$$p = -\frac{\partial F}{\partial V^*} = \frac{RT}{V^*}\frac{n}{N} = \frac{RT}{N}\nu.$$

This analysis shows that the existence of osmotic pressure can be deduced from the molecular-kinetic theory of heat, and that, at high dilution, according to this theory, equal

[2] A. Einstein, *Ann. d. Phys.* 11 (1903): 170.

numbers of solute molecules and suspended particles behave identically as regards osmotic pressure.

3. THEORY OF DIFFUSION OF SMALL SUSPENDED SPHERES

Suppose suspended particles are randomly distributed in a liquid. We will investigate their state of dynamic equilibrium under the assumption that a force K, which depends on the position but not on the time, acts on the individual particles. For the sake of simplicity, we shall assume that the force acts everywhere in the direction of the X-axis.

If the number of suspended particles per unit volume is ν, then in the case of thermodynamic equilibrium ν is a function of x such that the variation of the free energy vanishes for an arbitrary virtual displacement δx of the suspended substance. Thus

$$\delta F = \delta E - T\,\delta S = 0.$$

Let us assume that the liquid has a unit cross section perpendicular to the X-axis, and that it is bounded by the planes $x = 0$ and $x = l$. We then have

$$\delta E = -\int_0^l K\nu\,\delta x\,dx$$

and

$$\delta S = \int_0^l R\frac{\nu}{N}\frac{\partial \delta x}{\partial x}\,dx = -\frac{R}{N}\int_0^l \frac{\partial \nu}{\partial x}\,\delta x\,dx.$$

The required equilibrium condition is therefore

$$-K\nu + \frac{RT}{N}\frac{\partial \nu}{\partial x} = 0 \tag{1}$$

or

$$K\nu - \frac{\partial p}{\partial x} = 0.$$

The last equation asserts that the force K is equilibrated by the force of osmotic pressure.

We can use equation (1) to determine the diffusion coefficient of the suspended substance. We can look upon the dynamic equilibrium state considered here as a superposition of two processes proceeding in opposite directions, namely:

1. A motion of the suspended substance under the influence of the force K that acts on each suspended particle.
2. A process of diffusion, which is to be regarded as the result of the disordered motions of the particles produced by thermal molecular motion.

If the suspended particles have spherical form (where P is the radius of the sphere) and the coefficient of viscosity of the liquid is k, then the force K imparts to an individual particle the velocity[3]

$$\frac{K}{6\pi kP},$$

and

$$\frac{\nu K}{6\pi kP}$$

particles will pass through a unit area per unit time.

Further, if D denotes the diffusion coefficient of the suspended substance and μ the mass of a particle, then

$$-D\frac{\partial(\mu\nu)}{\partial x} \text{ grams}$$

[3]Cf., e.g., G. Kirchhoff, *Vorlesungen über Mechanik*, 26. Vorl., §4 (*Lectures on Mechanics*, Lecture 26, sec. 4).

or

$$-D\frac{\partial \nu}{\partial x}$$

particles will pass across a unit area per unit time as the result of diffusion. Since dynamic equilibrium prevails, we must have

$$\frac{\nu K}{6\pi kP} - D\frac{\partial \nu}{\partial x} = 0. \tag{2}$$

From the two conditions (1) and (2) found for dynamic equilibrium, we can calculate the diffusion coefficient. We get

$$D = \frac{RT}{N} \cdot \frac{1}{6\pi kP}.$$

Thus, except for universal constants and the absolute temperature, the diffusion coefficient of the suspended substance depends only on the viscosity of the liquid and on the size of the suspended particles.

4. ON THE DISORDERED MOTION OF PARTICLES SUSPENDED IN A LIQUID AND ITS RELATION TO DIFFUSION

We shall now turn to a closer examination of the disordered motions that arise from thermal molecular motion and give rise to the diffusion investigated in the last section.

Obviously, we must assume that each individual particle executes a motion that is independent of the motions of all the other particles; the motions of the same particle in different time intervals must also be considered as mutually independent processes, so long as we think of these time intervals as chosen not to be too small.

We now introduce a time interval τ, which is very small compared with observable time intervals but still large enough that the motions performed by a particle during two consecutive time intervals τ can be considered as mutually independent events.

Suppose, now, that a total of n suspended particles is present in a liquid. In a time interval τ, the X-coordinates of the individual particles will increase by Δ, where Δ has a different (positive or negative) value for each particle. A certain probability distribution law will hold for Δ: the number dn of particles experiencing a displacement that lies between Δ and $\Delta + d\Delta$ in the time interval τ will be expressed by an equation of the form

$$dn = n\varphi(\Delta)\,d\Delta,$$

where

$$\int_{-\infty}^{+\infty} \varphi(\Delta)\,d\Delta = 1,$$

and φ differs from zero only for very small values of Δ and satisfies the condition

$$\varphi(\Delta) = \varphi(-\Delta).$$

We will now investigate how the diffusion coefficient depends on φ, restricting ourselves again to the case where the number ν of particles per unit volume only depends on x and t.

Let $\nu = f(x, t)$ be the number of particles per unit volume; we calculate the distribution of particles at time $t + \tau$ from their distribution at time t. From the definition of the function $\varphi(\Delta)$ we can easily obtain the number of particles found at time $t + \tau$ between two planes perpendicular to the X-axis with abscissas x and $x + dx$. We get

$$f(x, t + \tau)\, dx = dx \cdot \int_{\Delta=-\infty}^{\Delta=+\infty} f(x + \Delta)\varphi(\Delta)\, d\Delta. \text{[1]}$$

But since τ is very small, we can put

$$f(x, t + \tau) = f(x, t) + \tau \frac{\partial f}{\partial t}.$$

Further, let us expand $f(x + \Delta, t)$ in powers of Δ :

$$f(x + \Delta, t) = f(x, t) + \Delta \, \frac{\partial f(x, t)}{\partial x} + \frac{\Delta^2}{2!} \, \frac{\partial^2 f(x, t)}{\partial x^2} \dots \text{ad inf.}$$

We can bring this expansion under the integral sign since only very small values of Δ contribute anything to the latter. We obtain

$$f + \frac{\partial f}{\partial t} \cdot \tau = f \cdot \int_{-\infty}^{+\infty} \varphi(\Delta)\, d\Delta + \frac{\partial f}{\partial x} \int_{-\infty}^{+\infty} \Delta\varphi(\Delta)\, d\Delta$$

$$+ \frac{\partial^2 f}{\partial x^2} \int_{-\infty}^{+\infty} \frac{\Delta^2}{2}\varphi(\Delta)\, d\Delta \dots$$

On the right-hand side, the second, fourth, etc., terms vanish since $\varphi(x) = \varphi(-x)$, while for the first, third, fifth, etc., terms, each successive term is very small compared with the one preceding it. From this equation, by taking into account that

$$\int_{-\infty}^{+\infty} \varphi(\Delta)\, d\Delta = 1,$$

and putting

$$\frac{1}{\tau} \int_{-\infty}^{+\infty} \frac{\Delta^2}{2}\varphi(\Delta)\, d\Delta = D,$$

and taking into account only the first and third terms of the right-hand side, we get

$$\frac{\partial f}{\partial t} = D \frac{\partial^2 f}{\partial x^2}. \tag{3}$$

This is the well-known differential equation for diffusion, and we recognize that D is the diffusion coefficient.

Another important point can be linked to this argument. We have assumed that the individual particles are all referred to the same coordinate system. However, this is not necessary since the motions of the individual particles are mutually independent. We will now refer the motion of each particle to a coordinate system whose origin coincides with the position of the center of gravity of the particle in question at time $t = 0$, with the difference that $f(x, t)dx$ now denotes the number of particles whose X-coordinate has *increased* between the times $t = 0$ and $t = t$ by a quantity that lies somewhere between x and $x + dx$. Thus, the function f varies according to equation (1) in this case as well.[2] Further, it is obvious that for $x \gtrless 0$ and $t = 0$ we must have

$$f(x, t) = 0 \text{ and } \int_{-\infty}^{+\infty} f(x, t)\, dx = n.$$

The problem, which coincides with the problem of diffusion outwards from a point (neglecting the interaction between the diffusing particles), is now completely determined mathematically; its solution is

$$f(x, t) = \frac{n}{\sqrt{4\pi D}} \frac{e^{-\frac{x^2}{4Dt}}}{\sqrt{t}}.$$

The probability distribution of the resulting displacements during an arbitrary time t is thus the same as the distribution of random errors, which was to be expected. What is important, however, is how the constant in the exponent is related to the diffusion coefficient. With the help of this equation we can now calculate the displacement λ_x in the direction of the X-axis that a particle experiences on the average, or, to be more precise, the root-mean-square displacement in the X-direction; it is

$$\lambda_x = \sqrt{x^2} = \sqrt{2Dt}.$$

The mean displacement is thus proportional to the square root of the time. It can easily be shown that the root mean square of the *total displacements* of the particles has the value $\lambda_x \sqrt{3}$.

5. FORMULA FOR THE MEAN DISPLACEMENT OF SUSPENDED PARTICLES. A NEW METHOD OF DETERMINING THE ACTUAL SIZE OF ATOMS

In section 3 we found the following value for the diffusion coefficient D of a substance suspended in a liquid in the form of small spheres of radius P :

$$D = \frac{RT}{N} \frac{1}{6\pi kP}.$$

Further, we found in section 4 that the mean value of the displacements of the particles in the X-direction at time t equals

$$\lambda_x = \sqrt{2Dt}.$$

By eliminating D, we get:

$$\lambda_x = \sqrt{t} \cdot \sqrt{\frac{RT}{N} \frac{1}{3\pi kP}}.$$

This equation shows how λ_x depends on T, k, and P.

We will now calculate how large λ_x is for one second if N is taken to be $6 \cdot 10^{23}$ in accordance with the results of the kinetic theory of gases; water at 17°C ($k = 1.35 \cdot 10^{-2}$) is chosen as the liquid,[3] and the diameter of the particles is 0.001 mm. We get

$$\lambda_x = 8 \cdot 10^{-5} \text{ cm} = 0.8 \text{ micron.}$$

Therefore, the mean displacement in one minute would be about 6 microns.

Conversely, the relation can be used to determine N. We obtain

$$N = \frac{t}{\lambda_x^2} \cdot \frac{RT}{3\pi kP}.$$

Let us hope that a researcher will soon succeed in solving the problem presented here, which is so important for the theory of heat.

(Annalen der Physik 17 [1905]: 549–560)

EDITORIAL NOTES

[1] f on the right-hand side is to be taken at the time t.
[2] Here eq. (1) should be eq. (3).
[3] The value of the viscosity of water is taken from paper 1, p. 65, and actually refers to water at temperature 9.5°C.

Part Three

✦

Einstein on the Theory of Relativity

$$k_y = \frac{d}{dt}\left\{\frac{mc^2}{\sqrt{1-\frac{q^2}{c^2}}}\right\} \quad \cdots \quad (27a)$$

$$E = \frac{mc^2}{\sqrt{1-\frac{q^2}{c^2}}} \quad \cdots \quad (28)$$

$$E = mc^2 + \frac{m}{2}q^2 + \cdots \quad (28')$$

Part of Einstein's 1912 manuscript on the special theory of relativity, showing the equation $E = mc^2$. (Courtesy of Hebrew University of Jerusalem)

Einstein was the first physicist to formulate clearly the new kinematical foundation for all of physics inherent in Lorentz's electron theory. This kinematics emerged in 1905 from his critical examination of the physical significance of the concepts of spatial and temporal intervals. The examination, based on a careful definition of the simultaneity of distant events, showed that the concept of a universal or absolute time, on which Newtonian kinematics is based, has to be abandoned; and that the Galilean transformations between the coordinates of two inertial frames of reference has to be replaced by a set of spatial and temporal transformations that agree formally with a set that Lorentz had introduced earlier with a quite different interpretation. Through the interpretation of these transformations as elements of a space-time symmetry group corresponding to the new kinematics, the special theory of relativity (as it later came to be called) provided physicists with a powerful guide in the search for new dynamical theories of fields and particles and gradually led to a deeper appreciation of the role of symmetry criteria in physics. The special theory of relativity also provided philosophers with abundant material for reflection on the new views of space and time. The special theory, like Newtonian mechanics, still assigns a privileged status to the class of inertial frames of reference. The attempt to generalize the theory to include gravitation led Einstein to formulate the equivalence principle in 1907. This was the first step in his search for a new theory of gravitation denying a privileged role to inertial frames, a theory that is now known as the general theory of relativity.

Einstein presented the special theory in paper 3, which is a landmark in the development of modern physics. In the first part of this paper Einstein presented the new kinematics, basing it on two postulates, the relativity principle and the principle of the constancy of the velocity of light. In the second part, he applied his kinematical results to the solution of a number of problems in the optics and electrodynamics of moving bodies. In paper 4, Einstein presents arguments for one of the most important consequences of the theory, the equivalence of mass and energy.

Strictly speaking, it is anachronistic to use the term "theory of relativity" in discussing Einstein's first papers on the subject. In them he referred to the "principle of relativity." Max Planck used the term "Relativtheorie" in 1906 to describe the Lorentz-Einstein equations of motion for the electron, and this expression continued to be used from time to time for several years. A. H. Bucherer seems to have been the first person to use the term "Relativitätstheorie" in the discussion following Planck's lecture. The term was used in an article by Paul Ehrenfest and adopted by Einstein in 1907 in his reply to this article. Although Einstein used the term from time to time thereafter, for several years he continued to employ "Relativitätsprinzip" in the titles of his articles. In 1910 the mathematician Felix Klein suggested the name "Invariantentheorie" (theory of invariants), but this suggestion does not seem to have been adopted by any physicist. In 1915 Einstein started to refer to his earlier work as "the special theory of relativity" to contrast it with his later "general theory."

IN HIS 1905 paper, as well as in his 1907 and 1909 reviews of the theory, Einstein described the theory of relativity as arising from a specific problem: the apparent conflict between the principle of relativity and the Maxwell-Lorentz theory of electrodynamics. While the relativity principle asserts the physical equivalence of all inertial frames of reference, the Maxwell-Lorentz theory implies the existence of a privileged inertial frame.

The principle of relativity originated in classical mechanics. Assuming Newton's laws of motion and central force interactions, it can be demonstrated that it is impossible to determine the state of motion of an inertial frame by means of mechanical experiments carried out within a closed system with center of mass at rest in this frame. This conclusion, well known and empirically well confirmed by the end of the nineteenth century, was sometimes called the principle of relative motion, or principle of relativity.

The introduction of velocity-dependent forces between charged particles led to doubts about the validity of the relativity principle for magnetic interactions. The wave theory of light appeared to invalidate the principle for optical phenomena. This theory seemed to require an all-pervading medium, the so-called luminiferous ether, to explain the propagation of light in the absence of ordinary matter. The assumption that the ether moves together with matter seems to be excluded by the phenomenon of aberration and by Fizeau's results on the velocity of light in moving media. If the ether is not dragged with matter, it should be possible to detect motion relative to a reference frame fixed in the ether by means of optical experiments. However, all attempts to detect the motion of the earth through the ether by optical experiments failed.

Maxwell's electromagnetic theory was intended to provide a unified explanation of electric, magnetic, and optical phenomena. With its advent, the question arose of the status of the principle of relativity for such phenomena. Does the principle follow from the fundamental equations of electrodynamics? The answer to this question depends on the form of Maxwell's equations postulated for bodies in motion. Hertz developed an electrodynamics of moving bodies, based on the assumption that the ether moves with matter, in which the relativity principle holds. In addition to its inability to account for the optical phenomena mentioned above, Hertz's theory was unable to explain several new electromagnetic phenomena, and it soon fell out of favor.

By the turn of this century, when Einstein started working on the electrodynamics of moving bodies, Lorentz's very successful version of Maxwell's theory had gained wide acceptance. Lorentz's electrodynamics is based on a microscopic theory that came to be known as the electron theory. The theory makes a sharp distinction between ordinary, ponderable matter and the ether. Ordinary matter is composed of finite-sized material particles, at least some of which are electrically charged. All of space, even those regions occupied by material particles, is pervaded by the ether, a medium with no mechanical properties, such as mass. The ether is the seat of all electric and magnetic fields. Matter only influences the ether through charged particles, which create these fields. The ether only acts on matter through the electric and magnetic forces that the fields exert on charged particles. By assuming such atoms of electricity ("electrons"), Lorentz's theory incorporates an important element of the pre-Maxwellian continental tradition into Maxwell's theory, from which it took the field equations.

The parts of the ether are assumed to be immobile relative to each other. Hence, Lorentz's ether defines a rigid reference frame, which is assumed to be inertial. It is in this frame that Maxwell's equations are valid; in other frames, the Galilei-transformed form of these equations hold. Hence it should be possible to detect the motion of the earth through the ether by suitably designed terrestrial electromagnetic or optical experiments. Lorentz was well aware of the failure of all attempts to detect the motion of the earth through the ether, in particular such sensitive optical attempts as the Michelson-Morley experiment, and attempted to explain this failure on the basis of his theory.

His basic approach to this problem in 1895 was to use the theorem of "corresponding states" in combination with the well-known contraction hypothesis. The theorem is essentially a calculational tool that sets up a correspondence between phenomena in moving systems and those in stationary systems by introducing transformed coordinates and fields. On this basis, Lorentz was able to account for the failure of most electromagnetic experiments to detect the motion of the earth through the ether. In 1904 he showed how to explain the failure of all such experiments by a generalization of his theorem. He introduced a set of transformations for the spatial and temporal coordinates (soon named the "Lorentz transformations" by Poincaré) and for the electric and magnetic field components, such that by using these transformations, Maxwell's equations, in the absence of charges, take the same form in all inertial frames. Lorentz's approach to the explanation of the failure of attempts to detect motion through the ether, thus, was to show that the basic equations of the electron theory, in spite of the fact that they single out the ether rest frame, can still explain this failure of all optical and electromagnetic attempts to detect the earth's motion through the ether.

PART THREE

Einstein's work was based on a new outlook on the problem. Instead of regarding the failure of electromagnetic and optical experiments to detect the earth's motion through the ether as something to be deduced from the electrodynamical equations, he took this failure as empirical evidence for the validity of the principle of relativity in electrodynamics and optics. Indeed, he asserted the universal validity of the principle, making it a criterion for the acceptability of any physical law. In this respect he gave the principle of relativity a role similar to that of the principles of thermodynamics, an example that he later stated helped to guide him. Rather than being deductions from other theories, such principles are taken as postulates for deductive chains of reasoning, resulting in the formulation of general criteria that all physical theories must satisfy.

Einstein now confronted the problem of making Maxwell-Lorentz electrodynamics compatible with the principle of relativity. He did so by means of a principle drawn from this very electrodynamics, the principle of the constancy of the velocity of light. That the velocity of light is independent of that of its source, and has a constant value in the ether rest frame, can be deduced from the Maxwell-Lorentz theory. Einstein dropped the ether from that theory and took the constancy of the velocity of light as a second postulate, supported by all the empirical evidence in favor of the Maxwell-Lorentz theory. When combined with the relativity principle, this leads to an apparently paradoxical conclusion: the velocity of light must be the same in all inertial frames. This result conflicts with the Newtonian law of addition of velocities, forcing a revision of the kinematical foundations underlying all of physics. Einstein showed that the simultaneity of distant events is only defined physically relative to a particular inertial frame, leading to kinematical transformations between the spatial and temporal coordinates of two inertial frames that agree formally with the transformations that Lorentz had introduced in 1904.

Einstein next considered the implications of the new kinematics for electrodynamics and mechanics. By eliminating the concept of the ether, he in effect asserted that electromagnetic fields do not require an underlying substratum. He showed that the Maxwell-Lorentz equations for empty space remain invariant in form under the new kinematical transformations when the transformation laws for the electric and magnetic fields are appropriately defined. He deduced appropriate transformation laws for charge densities and velocities from the requirement that Maxwell's equations remain invariant when convection currents are added. Finally, by assuming that Newton's equations hold for a charged particle at rest, he was able to use a kinematical transformation to deduce the equations of motion of a charged particle ("electron") with arbitrary velocity.

The problems connected with the formulation of an electrodynamics of moving bodies consistent with all experimental evidence were discussed frequently during the years Einstein was working on his theory. Statements similar to many of the individual points made in paper 3 occur in the contemporary literature, and Einstein may well have been familiar with some of the books and articles in which they are found. But his approach to the problem, leading to the peculiar combination of these ideas in his paper, is unique—particularly the recognition that a new kinematics of universal applicability is needed as the basis for a consistent approach to the electrodynamics of moving bodies.

EINSTEIN'S work on relativity grew out of his long-standing interest in the electrodynamics and optics of moving bodies His first scientific essay, written in 1895, discussed the propagation of light through the ether. The next year, as he later recalled, the following problem started to puzzle him: "If one were to pursue a light wave with the velocity of light, one would be confronted with a time-independent wave field. Such a thing doesn't seem to exist, however! This was the first childlike thought-experiment concerned with the special theory of relativity."[1]

By this time Einstein presumably was familiar with the principle of relativity in classical mechanics. While preparing for the ETH entrance examination in 1895, he had studied the German edition of Violle's textbook. Violle actually based his treatment of dynamics on the "principle of relative motions" together with the principle of inertia.

About 1898, Einstein started to study Maxwell's electromagnetic theory, apparently with the help of Drude's textbook. By 1899, after studying Hertz's papers on the subject, he was at work on the electrodynamics of moving bodies. He discussed this topic a number of times in letters to Mileva Marić between 1899 and 1901; once, on 27 March 1901, he referred to "our work on relative motion." In December 1901, Einstein also explained his ideas on the subject to Professor Alfred Kleiner of the University of Zurich, who encouraged him to publish them; but there is no evidence that Kleiner played a further role in the development of these ideas.

Einstein's comments show that in 1899 his viewpoint on electrodynamics was similar to that of Lorentz; but, aside from this similarity, there is no evidence that Einstein had yet read anything by Lorentz. Shortly afterward, Einstein designed an experiment to test the effect of the motion of bodies relative to the ether on the propagation of light; in 1901 he designed a second such experiment, but was unable to carry out either one. On 17 December 1901, he reported to Marić that he was at work on "a capital paper" on the electrodynamics of moving bodies, asserting his renewed conviction of the correctness of his "ideas on relative motion." His words may indicate that he already doubted whether motion with respect to the ether is experimentally detectable. Soon afterward he wrote that he intended to study Lorentz's theory in earnest.

There is direct or strong indirect contemporary evidence that, by 1902, Einstein had read or was reading works on electrodynamics and optics by Drude, Helmholtz, Hertz, Lorentz, Voigt, Wien, and Föppl. Comments in his letters on articles published in the *Annalen der Physik* between 1898 and 1901 indicate that during those years he looked at that journal regularly as well, and studied a number of articles in it. It is reasonable to suppose that he continued to do so between 1902 and 1905. During these years a number of significant articles on the electrodynamics and optics of moving bodies appeared in the *Annalen*. He cited several works published before 1905 in his later articles on relativity, and it is possible that he read one or more of these before 1905. Einstein also read extensively on the foundations of science. He later attributed great significance for his development of the theory of relativity to his reading of Hume, Mach, and Poincaré.

Belief in the reality of the ether was widespread at the turn of the century. However, Einstein was familiar with several works that questioned the certainty of its existence. Mill, in the course of a discussion of "the Hypothetical Method" in his *Logic*, gives a number of reasons for skepticism concerning "the prevailing hypothesis of a luminiferous ether."[2] Poincaré, in *La science et l'hypothèse*, raised the question of the existence of the ether, even if he offered no clear answer. Ostwald, in his *Lehrbuch der allgemeinen Chemie*, suggested that the ether hypothesis could be replaced by a purely energetic treatment of radiation.

Few contemporary documents throw any light on Einstein's work on electrodynamics between 1902 and 1905. On 22 January 1903, he wrote Michele Besso: "In the near future I want to deal with molecular forces in gases, and then make a comprehensive study of electron theory." On 5 December 1903, Einstein gave a talk to the Naturforschende Gesellschaft Bern on "The Theory of Electromagnetic Waves." By the time he wrote his friend Conrad Habicht in May or June 1905, the theory was practically complete: "The . . . paper is only a rough draft at this point, and is an electrodynamics of moving bodies, which employs a modification of the theory of space and time."

Later reminiscences by Einstein suggest several important elements in the development of his ideas on relativity before "On the Electrodynamics of Moving Bodies" was written that are not recorded in any known contemporary documents. In a letter to Erika Oppenheimer on 13 September 1932, he gave a general characterization of "the situation that led to setting up the theory of special relativity": "Mechanically all inertial systems are equivalent. In accordance with experience, this equivalence also extends to optics and electrodynamics. However, it did not appear that this equivalence could be attained in the theory of the latter. I soon reached the conviction that this had its basis in a deep incompleteness of the theoretical system. The desire to discover and overcome this generated a state of psychic tension in me that, after seven years of vain searching, was resolved by relativizing the concepts of time and length."

In 1952 he wrote: "My direct path to the special theory of relativity was mainly determined by the conviction that the electromotive force induced in a conductor moving in a magnetic field is nothing other than an electric field. But the result of Fizeau's experiment and the phenomenon of aberration also guided me."[3]

Beyond their well-known role as evidence against the assumption that the ether is completely carried along by moving matter, it is not clear what role the result of Fizeau's experiment and the phenomenon of aberration played in Einstein's thinking. Possibly its role depended on the fact that, in both cases, the observed effect only depends on the motion of matter (water in the first case, a star in the second) relative to the earth, and not on the presumed motion of the earth with respect to the ether.

In the case of electromagnetic induction, Einstein gave a more detailed account of its role. In 1920, he wrote: "In setting up the special theory of relativity, the following . . . idea about Faraday's electromagnetic induction played a guiding role. According to Faraday, relative motion of a magnet and a closed electric circuit induces an electric current in the latter. Whether the magnet is moved or the conductor doesn't matter; only the relative motion is significant. . . . The phenomena of electromagnetic induction . . . compelled me to postulate the principle of (special) relativity." In a footnote he added: "The difficulty to be overcome then lay in the constancy of the velocity of light in vacuum, which I first thought would have to be abandoned. Only after groping for years did I realize that the difficulty lay in the arbitrariness of the fundamental concepts of kinematics."[4]

His strong belief in the relativity principle and abandonment of "the constancy of the velocity of light in vacuum" led Einstein to explore the possibility of an emission theory of light. In such a theory, the velocity of light is only constant relative to that of its source, so it is clearly consistent with the relativity principle. Newton's corpuscular theory of light is an emission theory, and Einstein's search for such a theory may have been connected with his light-quantum hypothesis (see paper 5). On 25 April 1912, in a letter to Paul Ehrenfest commenting on Ritz's emission theory, Einstein referred to "Ritz's conception, which before the theory of relativity was also mine." He expanded on this remark on 20 June: "I knew that the principle of the constancy of the velocity of light was something quite independent of the relativity postulate, and I weighed which was more probable, the principle of the constancy of c [the speed of light], as required by Maxwell's equations, or the constancy of c exclusively for an observer located at the light source. I decided in favor of the former."

In 1924, Einstein described the sudden resolution of his dilemma: "After seven years of reflection in vain (1898–1905), the solution came to me suddenly with the thought that our concepts and laws of space and time can only claim validity insofar as they stand in a clear relation to our experiences; and that experience could very well lead to the alteration of these concepts and laws. By a revision of the concept of simultaneity into a more malleable form, I thus arrived at the special theory of relativity."[5]

In a talk at Kyoto University in 1922, Einstein is reported to have said that, after a year of struggle with the problem of how to reconcile Lorentz's theory with his ideas on relativity, he visited a friend one day to discuss the problem in detail with him. The next day Einstein said to his friend: "Thanks to you, I have completely solved my problem."[6] The friend was presumably Michele Besso, then his colleague at the Swiss Patent Office and the only person whose help is acknowledged in paper 3.

Work on this paper was apparently completed very rapidly after this. In March 1952 Einstein wrote to Carl Seelig that "between the conception of the idea for the special theory of relativity and the completion of the relevant publication, five or six weeks elapsed."

Einstein's comments suggest the following stages in his work on the theory of relativity:

1. He became convinced that, as is the case for mechanical phenomena, only the relative motions of ponderable bodies are significant in determining electromagnetic and optical phenomena; at some point, this conviction led him to abandon the concept of the ether.

2. He temporarily abandoned Lorentz's theory of electrodynamics, which appears to attach physical significance to absolute motion (i.e., motion with respect to empty space or the ether).

3. He explored the possibility of an alternative electrodynamical theory, which would justify the emission hypothesis about the constancy of the velocity of light relative to its source.

4. Abandoning such attempts, he reexamined Lorentz's theory, at some point focusing his concern on the conflict of his ideas on relative motion with a particular consequence of Lorentz's theory: the independence of velocity of light of the velocity of its source.

113

5. He recognized that this conflict involves previously tacitly accepted kinematical assumptions about temporal and spatial intervals, leading him to examine the meaning of the concept of the simultaneity of distant events.
6. He defined simultaneity physically, and constructed a new kinematical theory based on the relativity principle and the light principle, thus resolving the apparent conflict between them.

There have been a number of attempts at a detailed reconstruction of Einstein's development of the theory of relativity, attempts which often differ significantly in their conclusions. Such a reconstruction has to take into account other strands in Einstein's work at this time. In particular, by the time he wrote the relativity paper, he no longer regarded Maxwell's electromagnetic theory as universally valid, and had already proposed his light quantum hypothesis (see paper 5). He had also shown that the equipartition theorem, which his work on the foundations of thermodynamics convinced him is valid for the most general classical-mechanical system, combined with Maxwell's theory, leads to an incorrect law for black-body radiation (see section 1 of paper 5). Thus, he already had challenged the unlimited validity of both classical mechanics and of Maxwell's theory.

Einstein later recalled that, uncertain how to proceed in the search for better theories of the structure of matter and radiation, he became convinced that "only the discovery of a universal formal principle could lead . . . to assured results."[7] Such principles play a role analogous in this respect with the role played by the principles of thermodynamics. The theory of relativity is based on just such principles: even though suggested originally by specific mechanical and electromagnetic theories, the principles of relativity and of the constancy of the velocity of light are supported by empirical evidence that is independent of the validity of these theories.

ACCORDING to his sister's memoir, Einstein was anxious about whether his relativity paper would be accepted by the *Annalen der Physik*. After it was accepted, he eagerly anticipated an immediate reaction to its publication, even though he expected it to be critical. He was greatly disappointed when his paper was not even mentioned in the following issues of the *Annalen*. Sometime afterward, she recounts, he received a letter from Planck, requesting explanations of a few obscure points in the work. "After the long period of waiting, this was the first sign that his paper was being read at all. The happiness of the young scholar was all the greater, since acknowledgment of his accomplishment came from one of the greatest contemporary physicists. . . . At that time Planck's interest signified infinitely much for the morale of the young physicist."[8]

Planck and Einstein continued to correspond, and during the fall of 1905 Planck discussed Einstein's paper in the University of Berlin's physics colloquium. During the next few years, Planck wrote several papers developing further consequences of the relativity principle, and interested his assistant, Max Laue, and one of his students, Kurd von Mosengeil, in working on related problems. A few years later, Einstein paid tribute to Planck's role in promoting the theory of relativity: "The attention that this theory so quickly received from colleagues is surely to be ascribed in large part to the resoluteness and warmth with which he intervened for this theory."[9]

Other physicists also started to discuss Einstein's work in 1905 and 1906. Two months after it appeared, Kaufmann cited it in a preliminary report of his recent experiments on the mass of electrons in β-rays. The following year, in a fuller discussion of his results, while noting that the two theories yield the same equations of motion for the electron, he gave the first clear account of the basic theoretical difference between Lorentz's and Einstein's views. Drude, the editor of the *Annalen*, cited Einstein's paper in the second edition of his standard text on optics, as well as in an article on optics in the *Handbuch der Physik*. Wilhelm Röntgen wrote to Einstein asking for copies of his papers on electrodynamics, presumably in connection with a talk Röntgen was to give on the equations of motion of the electron. Sommerfeld, who heard the talk, soon read Einstein's work and was so impressed that he decided to give a colloquium on it. During 1907, Einstein was in correspondence about the theory with Planck, Laue, Wien, and Minkowski. In the same year, he was asked to write a review article on relativity, which appeared in Stark's *Jahrbuch der Radioaktivität* at the end of the year, and a major publishing house inquired about the possibility of a book on his research. A reference by Ehrenfest in 1907 to Einstein's theory as a "closed system" led Einstein to clarify his view of the nature of the theory. By 1908, the theory of relativity, though still controversial and often not clearly distinguished from Lorentz's electron theory, was a major topic of discussion among leading German-speaking physicists.

SINCE the theory of relativity grew out of Einstein's long-standing concern with electrodynamics, and his applications of the theory were primarily in this field, the theory was often looked upon as essentially another version of Lorentz's electron theory. Einstein soon felt the need to make clear the distinction between the kinematical results of the theory, deduced from the two principles of the theory, and the use of such kinematical results in the solution of problems in the optics and electrodynamics of moving bodies, the derivation of the equations of motion of a charged particle—or indeed in any physical theory. He pointed out that the postulates of the theory do not constitute a "closed system" but only a "heuristic principle, which considered by itself alone only contains assertions about rigid bodies, clocks, and light signals." Beyond such assertions, the theory could only establish "relations between otherwise apparently independent laws" of physics.[10]

A few months after first publishing the theory of relativity, Einstein discovered something that particularly intrigued him: the relation between inertial mass and energy. He wrote to Conrad Habicht during the summer of 1905: "One more consequence of the paper on electrodynamics has also occurred to me. The principle of relativity, in conjunction with Maxwell's equations, requires that mass be a direct measure of the energy contained in a body; light carries mass with it. A noticeable decrease of mass should occur in the case of radium. The argument is amusing and seductive; but for all I know the Lord might be laughing over it and leading me around by the nose."

The idea that inertial mass is associated with electromagnetic energy was often discussed before 1905. Around the turn of the century, it was suggested that all mechanical concepts might be derivable from those of electromagnetism. In particular, there were attempts to derive the entire inertial mass of the electron from the energy associated with its electromagnetic field. It was also proved that a radiation-filled container manifests an apparent inertial mass, which (if the mass of the container is neglected) is proportional to the energy of the enclosed radiation.

In paper 4, Einstein argued that, as a consequence of the relativity principle, inertial mass is associated with *all* forms of energy. He was only able to establish this result for a process involving the emission of electromagnetic radiation by a system, but argued that the result is independent of the mechanism by which the system loses energy. In addition, he was only able to show that a change in energy is associated with a change in inertial mass equal to the change in energy divided by c^2. His argument was criticized in 1907 by Planck, who presented his own argument to show that a transfer of heat is associated with a similarly related transfer of inertial mass.

Soon afterward, Stark attributed the discovery of the relation between mass and energy to Planck. Einstein wrote Stark on 17 February 1908: "I was rather disturbed that you do not acknowledge my priority with regard to the connection between inertial mass and energy." After receiving a conciliatory reply from Stark, acknowledging his priority, Einstein replied on 22 February, regretting his original testy reaction: "People, to whom it is granted to contribute something to the progress of science, should not allow pleasure in the fruits of their common work to be clouded by such matters."

Einstein returned to the relation between inertial mass and energy in 1906 and in 1907, giving more general arguments for their complete equivalence, but he did not achieve the complete generality to which he aspired. In his 1909 Salzburg talk, Einstein strongly emphasized that inertial mass is a property of all forms of energy, and therefore electromagnetic radiation must have mass. This conclusion strengthened Einstein's belief in the hypothesis that light quanta manifest particle-like properties.

In 1905, Einstein proposed a number of other experimentally testable consequences of his theory, in particular the equations of motion of the electron. The following year he suggested an experimental test of these equations employing cathode rays.[11]

In this paper, he also mentioned Kaufmann's experimental investigations of the motion of electrons in β-rays. Starting in 1901, Kaufmann had carried out a series of experiments on the deflection of β-rays by electric and magnetic fields. In 1905 he asserted that his recent experiments yielded data for the dependence of mass on velocity that were incompatible with the (identical) predictions of the Lorentz and Einstein theories. Kaufmann's work occasioned considerable discussion. Lorentz was disheartened by the apparent refutation of his theory. Planck subjected the experiment to a careful analysis, and concluded that Kaufmann's work could not be regarded as a definitive refutation of the Lorentz-Einstein predictions. Röntgen, one of the leading German experimentalists, is reported also to have felt that Kaufmann's results were not decisive, because his observations were not that accurate. In a 1907 review, Einstein discussed Kaufmann's results at some length, especially their apparent irreconcilability with the Lorentz-Einstein predictions. Commenting on a figure showing Kaufmann's results and the relativistic predictions, Einstein wrote: "Considering the difficulty of the experiment, one might be inclined to regard the agreement as satisfactory." However, he noted that the deviations were systematic and well outside Kaufmann's error limits. "Whether the systematic deviations are based upon a source of error not yet considered, or on lack of correspondence between the foundations of the theory of relativity and the facts, can only be decided with certainty when more manifold observational data are at hand."[12]

Although Einstein evidently accepted experiment as the ultimate arbiter of the fate of the theory, he was cautious about accepting Kaufmann's results as definitive, perhaps because of his familiarity with Planck's critical analysis of the experiments. What he found even more difficult to accept were alternative equations of motion for the electron that are based on what he regarded as arbitrary dynamical assumptions about the shape of a moving electron. While conceding that Kaufmann's data seemed to favor the theories of Abraham and Bucherer, Einstein concluded: "In my opinion, however, a rather small probability should be ascribed to these theories, since their fundamental assumptions about the mass of a moving electron are not supported by theoretical systems that embrace wider complexes of phenomena."[13]

This cautious attitude toward Kaufmann's results proved justified. During the following years, controversies over the interpretation of the experimental results prevented them from playing a decisive role in contemporary evaluations of the theory of relativity. Bestelmeyer carried out β-ray experiments generally regarded as inconclusive, while Bucherer's results favoring the Lorentz-Einstein equations were seriously questioned. Experiments using cathode rays, reported by several investigators starting in 1910, proved similarly inconclusive. Almost a decade elapsed until data supporting the relativistic predictions (the 1916 results of Guye and Lavanchy) were generally accepted.

EINSTEIN ON THE THEORY OF RELATIVITY

EDITORIAL NOTES

[1]Einstein, "Erinnerungen—Souvenirs," *Schweizerische Hochschulzeitung* 28 (*Sonderheft*) (1955): 146.

[2]John Stuart Mill, *A System of Logic Ratiocinative and Inductive: Being a Connected View of the Principles of Evidence and the Methods of Scientific Investigation*, 8th ed. (London: Longmans, Green, Reader and Dyer, 1872; 1st ed., 1843), vol. 2, pp. 12, 20, 23.

[3]Message by Einstein, prepared for R. S. Shankland to read at a celebration of the centennial of Michelson's birth, 19 December 1952, at Case Institute.

[4]Unpublished manuscript, entitled "Fundamental Ideas and Methods of the Theory of Relativity, Presented as It Developed," copy in Einstein Editorial Archive, Boston University.

[5]Recording, transcribed in Friedrich Herneck, "Zwei Tondokumente Einsteins zur Relativitätstheorie," *Forschungen und Fortschritte* 40 (1966): 134.

[6]See the report of Einstein's talk, given on 14 December 1922, in Jun Ishiwara, *Einstein Kyôzyu-Kôen-roku* (Tokyo: Kabushika Kaisha, 1971), pp. 78–88.

[7]Einstein, *Autobiographical Notes*, Paul Arthur Schilpp, trans. and ed. (La Salle, Ill.: Open Court, 1979), p. 48.

[8]Maja Winteler-Einstein, "Albert Einstein: Beitrag für sein Lebensbild," typescript, pp. 23–24, Einstein Editorial Archive, Boston University.

[9]*Die Naturwissenschaften* 1 (1913): 1079, reprinted in *Collected Papers*, vol. 4, doc. 23, pp. 561–563.

[10]Einstein, "Comments on the Note of Mr. Paul Ehrenfest: 'The Translatory Motion of Deformable Electrons and the Area Law,' " *Collected Papers*, vol. 2, doc. 44, pp. 410-412.

[11]*Annalen der Physik* 21 (1906): 583–586, reprinted in *Collected Papers*, vol. 2, doc. 36, pp. 368–371.

[12]*Jahrbuch der Radioaktivität und Elektronik* 4 (1907): 433–462, citation on pp. 437–439, reprinted in *Collected Papers*, vol. 2, doc. 47, pp. 433–484.

[13]Ibid.

PAPER 3

✦

On the Electrodynamics
of Moving Bodies

IT IS WELL KNOWN that Maxwell's electrodynamics—as usually understood at present—when applied to moving bodies, leads to asymmetries that do not seem to be inherent in the phenomena. Take, for example, the electrodynamic interaction between a magnet and a conductor. The observable phenomenon here depends only on the relative motion of conductor and magnet, whereas the customary view draws a sharp distinction between the two cases, in which either the one or the other of the two bodies is in motion. For if the magnet is in motion and the conductor is at rest, an electric field with a definite energy value results in the vicinity of the magnet that produces a current wherever parts of the conductor are located. But if the magnet is at rest while the conductor is moving, no electric field results in the vicinity of the magnet, but rather an electromotive force in the conductor, to which no energy per se corresponds, but which, assuming an equality of relative motion in the two cases, gives rise to electric currents of the same magnitude and the same course as those produced by the electric forces in the former case.

Examples of this sort, together with the unsuccessful attempts to detect a motion of the earth relative to the "light medium," lead to the conjecture that not only the phenomena of mechanics but also those of electrodynamics have no properties that correspond to the concept of absolute rest. Rather, the same laws of electrodynamics and optics will be valid[1] for all coordinate systems in which the equations of mechanics hold, as has already been shown for quantities of the first order. We shall raise this conjecture (whose content will hereafter be called "the principle of relativity") to the status of a postulate and shall also introduce another postulate, which is only seemingly incompatible with it, namely that light always propagates in empty space with a definite velocity V that is independent of the state of motion of the emitting body. These two postulates suffice for the attainment of a simple and consistent electrodynamics of moving bodies based on Maxwell's theory for bodies at rest. The introduction of a "light ether" will prove to be superfluous, inasmuch as the view to be developed here will not require a "space at absolute rest" endowed with special properties, nor assign a velocity vector to a point of empty space where electromagnetic processes are taking place.

Like all electrodynamics, the theory to be developed here is based on the kinematics of a rigid body, since the assertions of any such theory have to do with the relations among rigid bodies (coordinate systems), clocks, and electromagnetic processes. Insufficient regard for this circumstance is at the root of the difficulties with which the electrodynamics of moving bodies currently has to contend.

A. Kinematic Part

1. Definition of Simultaneity

Consider a coordinate system in which Newton's mechanical equations are valid. To distinguish this system verbally from those to be introduced later, and to make our presentation more precise, we will call it the "rest system."

If a particle is at rest relative to this coordinate system, its position relative to the latter can be determined by means of rigid measuring rods using the methods of Euclidean geometry and expressed in Cartesian coordinates.

If we want to describe the *motion* of a particle, we give the values of its coordinates as functions of time. However, we must keep in mind that a mathematical description of this kind only has physical meaning if we are already clear as to what we understand here by "time." We have to bear in mind that all our judgments involving time are always judgments about *simultaneous events*. If, for example, I say that "the train arrives here at 7 o'clock," that means, more or less, "the pointing of the small hand of my watch to 7 and the arrival of the train are simultaneous events."[1]

It might seem that all difficulties involved in the definition of "time" could be overcome by my substituting "position of the small hand of my watch" for "time." Such a definition is indeed sufficient if a time is to be defined exclusively for the place at which the watch is located; but the definition is no longer satisfactory when series of events occurring at different locations have to be linked temporally, or—what

---

[1] We shall not discuss here the imprecision inherent in the concept of simultaneity of two events taking place at (approximately) the same location, which can be removed only by an abstraction.

amounts to the same thing—when events occurring at places remote from the clock have to be evaluated temporally.

To be sure, we could content ourselves with evaluating the time of events by stationing an observer with a clock at the origin of the coordinates who assigns to an event to be evaluated the corresponding position of the hands of the clock when a light signal from that event reaches him through empty space. However, we know from experience that such a coordination has the drawback of not being independent of the position of the observer with the clock. We reach a far more practical arrangement by the following argument.

If there is a clock at point A in space, then an observer located at A can evaluate the time of events in the immediate vicinity of A by finding the positions of the hands of the clock that are simultaneous with these events. If there is another clock at point B that in all respects resembles the one at A, then the time of events in the immediate vicinity of B can be evaluated by an observer at B. But it is not possible to compare the time of an event at A with one at B without a further stipulation. So far we have defined only an "A-time" and a "B-time," but not a common "time" for A and B. The latter can now be determined by establishing *by definition* that the "time" required for light to travel from A to B is equal to the "time" it requires to travel from B to A. For, suppose a ray of light leaves from A for B at "A-time" t_A, is reflected from B toward A at "B-time" t_B, and arrives back at A at "A-time" t'_A. The two clocks are synchronous by definition if

$$t_B - t_A = t'_A - t_B.$$

We assume that it is possible for this definition of synchronism to be free of contradictions, and to be so for arbitrarily many points, and therefore that the following relations are generally valid:

1. If the clock at B runs synchronously with the clock at A, the clock at A runs synchronously with the clock at B.
2. If the clock at A runs synchronously with the clock at B as well as with the clock at C, then the clocks at B and C also run synchronously relative to each other.

By means of certain (imagined) physical experiments, we have established what is to be understood by synchronous clocks at rest relative to each other and located at different places, and thereby obviously arrived at definitions of "synchronous" and "time." The "time" of an event is the reading obtained simultaneously from a clock at rest that is located at the place of the event, which for all time determinations runs synchronously with a specified clock at rest, and indeed with the specified clock.

Based on experience, we further stipulate that the quantity

$$\frac{2\,\overline{AB}}{t'_A - t_A} = V$$

be a universal constant (the velocity of light in empty space).

It is essential that we have defined time by means of clocks at rest in the rest system; because the time just defined is related to the system at rest, we call it "the time of the rest system."

2. On the Relativity of Lengths and Times

The following considerations are based on the principle of relativity and the principle of the constancy of the velocity of light. We define these two principles as follows:

1. If two coordinate systems are in uniform parallel translational motion relative to each other, the laws according to which the states of a physical system change do not depend on which of the two systems these changes are related to.

2. Every light ray moves in the "rest" coordinate system with a fixed velocity V, independently of whether this ray of light is emitted by a body at rest or in motion. Hence,

$$\text{velocity} = \frac{\text{light path}}{\text{time interval}},$$

where "time interval" should be understood in the sense of the definition given in section 1.

Take a rigid rod at rest; let its length, measured by a measuring rod that is also at rest, be l. Now imagine the axis of the rod placed along the X-axis of the rest coordinate system, and the rod then set into uniform parallel translational motion (with velocity v) along the X-axis in the direction of increasing x. We now inquire about the length of the *moving* rod, which we imagine to be ascertained by the following two operations:

a. The observer moves together with the aforementioned measuring rod and the rigid rod to be measured, and measures the length of the rod by laying out the measuring rod in the same way as if the rod to be measured, the observer, and the measuring rod were all at rest.

b. Using clocks at rest and synchronous in the rest system as outlined in section 1, the observer determines at which points of the rest system the beginning and end of the rod to be measured are located at some given time t. The distance between these two points, measured with the rod used before— but now at rest—is also a length that we can call the "length of the rod."

According to the principle of relativity, the length determined by operation (a), which we shall call "the length of the rod in the moving system," must equal the length l of the rod at rest.

The length determined using operation (b), which we shall call "the length of the (moving) rod in the rest system," will be determined on the basis of our two principles, and we shall find that it differs from l.

Current kinematics tacitly assumes that the lengths determined by the above two operations are exactly equal to each other, or, in other words, that at the time t a moving rigid body is totally replaceable, in geometric respects, by the *same* body when it is *at rest* in a particular position.

Further, we imagine the two ends (A and B) of the rod equipped with clocks that are synchronous with the clocks of the rest system, i.e., whose readings always correspond to the "time of the system at rest" at the locations the clocks happen to occupy; hence, these clocks are "synchronous in the rest system."

We further imagine that each clock has an observer comoving with it, and that these observers apply to the two clocks the criterion for the synchronous rate of two clocks formulated in section 1. Let a ray of light start out from A at time[2] t_A; it is reflected from B at time t_B, and arrives back at A at time t'_A. Taking into account the principle of the constancy of the velocity of light, we find that

$$t_B - t_A = \frac{r_{AB}}{V - v}$$

[2] "Time" here means both "time of the system at rest" and "the position of the hands of the moving clock located at the place in question."

and

$$t'_A - t_B = \frac{r_{AB}}{V + v},$$

where r_{AB} denotes the length of the moving rod, measured in the rest system. Observers co-moving with the rod would thus find that the two clocks do not run synchronously, while observers in the system at rest would declare them to be running synchronously.

Thus we see that we cannot ascribe *absolute* meaning to the concept of simultaneity; instead, two events that are simultaneous when observed from some particular coordinate system can no longer be considered simultaneous when observed from a system that is moving relative to that system.

3. Theory of Transformations of Coordinate and Time from the Rest System to a System in Uniform Translational Motion Relative to It

Let there be two coordinate systems in the "rest" space, i.e., two systems of three mutually perpendicular rigid material lines originating from one point. Let the X-axes of the two systems coincide, and their Y- and Z-axes be respectively parallel. Each system shall be supplied with a rigid measuring rod and a number of clocks, and let both measuring rods and all the clocks of the two systems be exactly alike.

Now, put the origin of one of the two systems, say k, in a state of motion with (constant) velocity v in the direction of increasing x of the other system (K), which remains at rest; and let this new velocity be imparted to k's coordinate axes, its corresponding measuring rod, and its clocks. To each time t of the rest system K, there corresponds a definite location of the axes of the moving system. For reasons of symmetry we are justified to assume that the motion of k can be such that at time t ("t" always denotes a time of the rest system) the axes of the moving system are parallel to the axes of the rest system.

We now imagine space to be measured out from both the rest system K using the measuring rod at rest, and from the moving system k using the measuring rod moving along with it, and that coordinates x, y, z and ξ, η, ζ respectively are obtained in this way. Further, by means of the clocks at rest in the rest system, and using light signals as described in section 1, we determine the time t of the rest system for all the points where there are clocks. In a similar manner, by again applying the method of light signals described in section 1, we determine the time τ of the moving system, for all points of this moving system at which there are clocks at rest relative to this system.

To every set of values x, y, z, t which completely determines the place and time of an event in the rest system, there corresponds a set of values ξ, η, ζ, τ that fixes this event relative to the system k, and the problem to be solved now is to find the system of equations that connects these quantities.

First of all, it is clear that these equations must be *linear* because of the properties of homogeneity that we attribute to space and time.

If we put $x' = x - vt$, then it is clear that a point at rest in the system k has a definite, time-independent set of values x', y, z belonging to it. We first determine τ as a function of x', y, z, and t. To this end, we must express in equations that τ is in fact the aggregate of readings of clocks at rest in system k, synchronized according to the rule given in section 1.

PAPER 3

Suppose that at time τ_0, a light ray is sent from the origin of the system k along the X-axis to x' and reflected from there toward the origin at time τ_1, arriving there at time τ_2; we then must have

$$\frac{1}{2}(\tau_0 + \tau_2) = \tau_1,$$

or, including the arguments of the function τ and applying the principle of the constancy of the velocity of light in the rest system,

$$\frac{1}{2}\left[\tau(0,0,0,t) + \tau\left(0,0,0,\left\{t + \frac{x'}{V-v} + \frac{x'}{V+v}\right\}\right)\right]$$
$$= \tau\left(x',0,0,t + \frac{x'}{V-v}\right).$$

From this we get, letting x' be infinitesimally small,

$$\frac{1}{2}\left(\frac{1}{V-v} + \frac{1}{V+v}\right)\frac{\partial \tau}{\partial t} = \frac{\partial \tau}{\partial x'} + \frac{1}{V-v}\frac{\partial \tau}{\partial t},$$

or

$$\frac{\partial \tau}{\partial x'} + \frac{v}{V^2-v^2}\frac{\partial \tau}{\partial t} = 0.$$

It should be noted that, instead of the coordinate origin, we could have chosen any other point as the origin of the light ray, and therefore the equation just derived holds for all values of x', y, z.

Analogous reasoning—applied to the $H^{[2]}$ and Z axes—yields, remembering that light always propagates along these axes with the velocity $\sqrt{V^2 - v^2}$ when observed from the rest system,

$$\frac{\partial \tau}{\partial y} = 0$$

$$\frac{\partial \tau}{\partial z} = 0.$$

These equations yield, since τ is a *linear* function,

$$\tau = a\left(t - \frac{v}{V^2 - v^2} x'\right),$$

where a is a function $\varphi(v)$ as yet unknown, and where we assume for brevity that at the origin of k we have $t = 0$ when $\tau = 0$.

Using this result, we can easily determine the quantities ξ, η, ζ by expressing in equations that (as demanded by the principle of the constancy of the velocity of light in conjunction with the principle of relativity) light also propagates with velocity V when measured in the moving system. For a light ray emitted at time $\tau = 0$ in the direction of increasing ξ, we have

$$\xi = V\tau,$$

or

$$\xi = aV\left(t - \frac{v}{V^2 - v^2} x'\right).$$

But as measured in the rest system, the light ray propagates with velocity $V - v$ relative to the origin of k, so that

$$\frac{x'}{V - v} = t.$$

Substituting this value of t in the equation for ξ, we obtain

$$\xi = a \frac{V^2}{V^2 - v^2} x'.$$

Analogously, by considering light rays moving along the two other axes, we get

$$\eta = V\tau = aV\left(t - \frac{v}{V^2 - v^2} x'\right),$$

133

where

$$\frac{y}{\sqrt{V^2 - v^2}} = t; \qquad x' = 0;$$

hence

$$\eta = a\frac{V}{\sqrt{V^2 - v^2}}\ y$$

and

$$\zeta = a\frac{V}{\sqrt{V^2 - v^2}}\ z.$$

If we substitute for x' its value, we obtain

$$\tau = \varphi(v)\,\beta\!\left(t - \frac{v}{V^2}\,x\right),$$

$$\xi = \varphi(v)\,\beta(x - vt),$$

$$\eta = \varphi(v)y,$$

$$\zeta = \varphi(v)z,$$

where

$$\beta = \frac{1}{\sqrt{1 - \left(\dfrac{v}{V}\right)^2}}$$

and φ is an as yet unknown function of v. If no assumptions are made regarding the initial position of the moving system and the zero point of τ, then a constant must be added to the right-hand sides of these equations.

Now we have to prove that, measured in the moving system, every light ray propagates with the velocity V, if it does so, as we have assumed, in the rest system; for we have not yet proved that the principle of the constancy of the velocity of light is compatible with the relativity principle.

Suppose that at time $t = \tau = 0$ a spherical wave is emitted from the coordinate origin, which at that time is common to both systems, and that this wave propagates in the system K with the velocity V. Hence, if (x, y, z) is a point reached by this wave, we have

$$x^2 + y^2 + z^2 = V^2 t^2.$$

We transform this equation using our transformation equations and, after a simple calculation, obtain

$$\xi^2 + \eta^2 + \zeta^2 = V^2 \tau^2.$$

Thus, our wave is also a spherical wave with propagation velocity V when it is observed in the moving system. This proves that our two fundamental principles are compatible.[3]

The transformation equations we have derived also contain an unknown function φ of v, which we now wish to determine.

To this end we introduce a third coordinate system K', which, relative to the system k, is in parallel-translational motion, parallel to the axis Ξ,[4] such that its origin moves along the Ξ-axis with velocity $-v$. Let all three coordinate origins coincide at time $t = 0$, and let the time t' of system K' equal zero at $t = x = y = z = 0$. We denote the coordinates measured in the system K' by x', y', z' and, by twofold application of our transformation equations, we get

$$t' = \varphi(-v)\beta(-v)\left\{\tau + \frac{v}{V^2}\xi\right\} = \varphi(v)\varphi(-v)t,$$

$$x' = \varphi(-v)\beta(-v)\{\xi + v\tau\} \qquad = \varphi(v)\varphi(-v)x,$$

$$y' = \varphi(-v)\eta \qquad\qquad\qquad = \varphi(v)\varphi(-v)y,$$

$$z' = \varphi(-v)\zeta \qquad\qquad\qquad = \varphi(v)\varphi(-v)z.$$

Since the relations between x', y', z' and x, y, z do not contain the time t, the systems K and K' are at rest relative to each other, and it is clear that the transformation from K to K' must be the identity transformation. Hence,

$$\varphi(v)\varphi(-v) = 1.$$

Let us now explore the meaning of $\varphi(v)$. We shall focus on that portion of the H-axis of the system k that lies between $\xi = 0,\ \eta = 0,\ \zeta = 0$, and $\xi = 0,\ \eta = l,\ \zeta = 0$. This portion of the H-axis is a rod that, relative to the system K, moves perpendicular to its axis with a velocity v and its ends have coordinates in K:

$$x_1 = vt, \quad y_1 = \frac{l}{\varphi(v)}, \quad z_1 = 0$$

and

$$x_2 = vt, \quad y_2 = 0, \quad z_2 = 0.$$

The length of the rod, measured in K, is thus $l/\varphi(v)$; this gives us the meaning of the function φ. For reasons of symmetry, it is now evident that the length of a rod measured in the rest system and moving perpendicular to its axis can depend only on its velocity and not on the direction and sense of its motion. Thus, the length of the moving rod measured in the rest system does not change if v is replaced by $-v$. From this we conclude:

$$\frac{l}{\varphi(v)} = \frac{l}{\varphi(-v)},$$

or

$$\varphi(v) = \varphi(-v).$$

From this relation and the one found earlier it follows that $\varphi(v) = 1$, so that the transformation equations obtained become

$$\tau = \beta\left(t - \frac{v}{V^2}\,x\right),$$

$$\xi = \beta(x - vt),$$

$$\eta = y,$$

$$\zeta = z,$$

where

$$\beta = \frac{1}{\sqrt{1 - \left(\dfrac{v}{V}\right)^2}}.$$

4. The Physical Meaning of the Equations Obtained as Concerns Moving Rigid Bodies and Moving Clocks

We consider a rigid sphere[3] of radius R that is at rest relative to the moving system k and whose center lies at the origin of k. The equation of the surface of this sphere, which moves with velocity v relative to the system k, is

$$\xi^2 + \eta^2 + \zeta^2 = R^2.$$

Expressed in terms of x, y, z, the equation of this surface at time $t = 0$ is

$$\frac{x^2}{\left(\sqrt{1 - \left(\dfrac{v}{V}\right)^2}\right)^2} + y^2 + z^2 = R^2.$$

A rigid body that has a spherical shape when measured at rest has, when in motion—considered from the rest

---

[3] I.e., a body that has a spherical shape when examined at rest.

system—the shape of an ellipsoid of revolution with axes

$$R\sqrt{1-\left(\frac{v}{V}\right)^2}, R, R.$$

Thus, while the Y and Z dimensions of the sphere (and hence also of every rigid body, whatever its shape) do not appear to be altered by motion, the X dimension appears to be contracted in the ratio $1 : \sqrt{1-(v/V)^2}$, thus the greater the value of v, the greater the contraction. For $v = V$, all moving objects—considered from the "rest" system—shrink into plane structures. For superluminal velocities our considerations become meaningless; as we shall see from later considerations, in our theory the velocity of light physically plays the role of infinitely great velocities.

It is clear that the same results apply for bodies at rest in the "rest" system when considered from a uniformly moving system.

We further imagine one of the clocks that is able to indicate time t when at rest relative to the rest system and time τ when at rest relative to the moving system to be placed at the origin of k and set such that it indicates the time τ. What is the rate of this clock when considered from the rest system?

The quantities x, t, and τ that refer to the position of this clock obviously satisfy the equations

$$\tau = \frac{1}{\sqrt{1-\left(\frac{v}{V}\right)^2}}\left(t - \frac{v}{V^2}x\right)$$

and

$$x = vt.$$

We thus have

$$\tau = t\sqrt{1 - \left(\frac{v}{V}\right)^2} = t - \left(1 - \sqrt{1 - \left(\frac{v}{V}\right)^2}\right)t,$$

from which it follows that the reading of the clock considered from the rest system lags behind each second by $(1 - \sqrt{1 - (v/V)^2})$ sec or, up to quantities of the fourth and higher order, by $\frac{1}{2}(v/V)^2$ sec.

This yields the following peculiar consequence: If at the points A and B of K there are clocks at rest that, considered from the rest system, are running synchronously, and if the clock at A is transported to B along the connecting line with velocity v, then upon arrival of this clock at B the two clocks will no longer be running synchronously; instead, the clock that has been transported from A to B will lag $\frac{1}{2}tv^2/V^2$ sec (up to quantities of the fourth and higher orders) behind the clock that has been in B from the outset, where t is the time needed by the clock to travel from A to B.

We see at once that this result holds even when the clock moves from A to B along any arbitrary polygonal line, and even when the points A and B coincide.[5]

If we assume that the result proved for a polygonal line holds also for a continuously curved line, then we arrive at the following result: If there are two synchronously running clocks at A, and one of them is moved along a closed curve with constant velocity until it has returned to A, which takes, say, t sec, then, on its arrival at A, this clock will lag $\frac{1}{2}t(v/V)^2$ sec behind the clock that has not been moved. From this we conclude that a balance-wheel clock[6] located at the Earth's equator must, under otherwise identical conditions, run more slowly by a very small amount than an absolutely identical clock located at one of the Earth's poles.

5. The Addition Theorem for Velocities

In the system k moving with velocity v along the X-axis of the system K, let a point move in accordance with the equations

$$\xi = w_\xi \tau,$$

$$\eta = w_\eta \tau,$$

$$\zeta = 0,$$

where w_ξ and w_η denote constants.

We seek the motion of the point relative to the system K. Introducing the quantities x, y, z, t into the equations of motion of the point by means of the transformation equations derived in section 3, we obtain

$$x = \frac{w_\xi + v}{1 + \dfrac{vw_\xi}{V^2}}\, t,$$

$$y = \frac{\sqrt{1 - \left(\dfrac{v}{V}\right)^2}}{1 + \dfrac{vw_\xi}{V^2}}\, w_\eta t,$$

$$z = 0.$$

Thus, according to our theory, the vector addition for velocities holds only to first approximation. Let

$$U^2 = \left(\frac{dx}{dt}\right)^2 + \left(\frac{dy}{dt}\right)^2,$$

$$w^2 = w_\xi^2 + w_\eta^2$$

and

$$\alpha = \arctan \frac{w_y}{w_x}; [7]$$

α is then to be considered as the angle between the velocities v and w. After a simple calculation we obtain

$$U = \frac{\sqrt{(v^2 + w^2 + 2vw \cos \alpha) - \left(\dfrac{vw \sin \alpha}{V}\right)^2}}{1 + \dfrac{vw \cos \alpha}{V^2}}.$$

It is worth noting that v and w enter into the expression for the resultant velocity in a symmetrical manner. If w also has the direction of the X-axis (Ξ-axis), we get

$$U = \frac{v + w}{1 + \dfrac{vw}{V^2}}.$$

It follows from this equation that the composition of two velocities that are smaller than V always results in a velocity that is smaller than V. For if we set $v = V - \kappa$, and $w = V - \lambda$, where κ and λ are positive and smaller than V, then

$$U = V \frac{2V - \kappa - \lambda}{2V - \kappa - \lambda + \dfrac{\kappa\lambda}{V}} < V.$$

It also follows that the velocity of light V cannot be changed by compounding it with a "subluminal velocity." For this case we get

$$U = \frac{V + w}{1 + \dfrac{w}{V}} = V.$$

In the case where v and w have the same direction, the formula for U could also have been obtained by compounding two transformations according to section 3. If in addition to the systems K and k, occurring in section 3, we introduce a third coordinate system k', which moves parallel to k and whose origin moves with velocity w along the Ξ-axis, we obtain equations between the quantities x, y, z, t and the corresponding quantities of k' that differ from those found in section 3 only insofar as "v" is replaced by the quantity

$$\frac{v + w}{1 + \frac{vw}{V^2}};$$

from this we see that such parallel transformations form a group—as indeed they must.

We have now derived the required laws of the kinematics corresponding to our two principles, and proceed to their application to electrodynamics.

B. Electrodynamic Part

6. Transformation of the Maxwell-Hertz Equations for Empty Space. On the Nature of the Electromotive Forces Arising Due to Motion in a Magnetic Field

Let the Maxwell-Hertz equations for empty space be valid for the rest system K, so that we have

$$\frac{1}{V}\frac{\partial X}{\partial t} = \frac{\partial N}{\partial y} - \frac{\partial M}{\partial z}, \qquad \frac{1}{V}\frac{\partial L}{\partial t} = \frac{\partial Y}{\partial z} - \frac{\partial Z}{\partial y},$$

$$\frac{1}{V}\frac{\partial Y}{\partial t} = \frac{\partial L}{\partial z} - \frac{\partial N}{\partial x}, \qquad \frac{1}{V}\frac{\partial M}{\partial t} = \frac{\partial Z}{\partial x} - \frac{\partial X}{\partial z},$$

$$\frac{1}{V}\frac{\partial Z}{\partial t} = \frac{\partial M}{\partial x} - \frac{\partial L}{\partial y}, \qquad \frac{1}{V}\frac{\partial N}{\partial t} = \frac{\partial X}{\partial y} - \frac{\partial Y}{\partial x},$$

where (X, Y, Z) denotes the electric force vector and (L, M, N) the magnetic force vector.

If we apply the transformations derived in section 3 to these equations, in order to relate the electromagnetic processes to the coordinate system moving with velocity v introduced there, we obtain the following equations:

$$\frac{1}{V}\frac{\partial X}{\partial \tau} = \frac{\partial \beta \left(N - \frac{v}{V}Y\right)}{\partial \eta} - \frac{\partial \beta \left(M + \frac{v}{V}Z\right)}{\partial \zeta},$$

$$\frac{1}{V}\frac{\partial \beta \left(Y - \frac{v}{V}N\right)}{\partial \tau} = \frac{\partial L}{\partial \zeta} - \frac{\partial \beta \left(N - \frac{v}{V}Y\right)}{\partial \xi},$$

$$\frac{1}{V}\frac{\partial \beta \left(Z + \frac{v}{V}M\right)}{\partial \tau} = \frac{\partial \beta \left(M + \frac{v}{V}Z\right)}{\partial \xi} - \frac{\partial L}{\partial \eta},$$

$$\frac{1}{V}\frac{\partial L}{\partial \tau} = \frac{\partial \beta \left(Y - \frac{v}{V}N\right)}{\partial \zeta} - \frac{\partial \beta \left(Z + \frac{v}{V}M\right)}{\partial \eta},$$

$$\frac{1}{V}\frac{\partial \beta \left(M + \frac{v}{V}Z\right)}{\partial \tau} = \frac{\partial \beta \left(Z + \frac{v}{V}M\right)}{\partial \xi} - \frac{\partial X}{\partial \zeta},$$

$$\frac{1}{V}\frac{\partial \beta \left(N - \frac{v}{V}Y\right)}{\partial \tau} = \frac{\partial X}{\partial \eta} - \frac{\partial \beta \left(Y - \frac{v}{V}N\right)}{\partial \xi},$$

where
$$\beta = \frac{1}{\sqrt{1 - \left(\frac{v}{V}\right)^2}}.$$

The relativity principle requires that the Maxwell-Hertz equations for empty space also be valid in the system k if they are valid in the system K, i.e, that the electric and magnetic force vectors—(X', Y', Z') and (L', M', N')—of the moving system k, which are defined in this system by their ponderomotive effects on electric and magnetic charges, respectively, satisfy the equations

$$\frac{1}{V}\frac{\partial X'}{\partial \tau} = \frac{\partial N'}{\partial \eta} - \frac{\partial M'}{\partial \zeta}, \qquad \frac{1}{V}\frac{\partial L'}{\partial \tau} = \frac{\partial Y'}{\partial \zeta} - \frac{\partial Z'}{\partial \eta},$$

$$\frac{1}{V}\frac{\partial Y'}{\partial \tau} = \frac{\partial L'}{\partial \zeta} - \frac{\partial N'}{\partial \xi}, \qquad \frac{1}{V}\frac{\partial M'}{\partial \tau} = \frac{\partial Z'}{\partial \xi} - \frac{\partial X'}{\partial \zeta},$$

$$\frac{1}{V}\frac{\partial Z'}{\partial \tau} = \frac{\partial M'}{\partial \xi} - \frac{\partial L'}{\partial \eta}, \qquad \frac{1}{V}\frac{\partial N'}{\partial \tau} = \frac{\partial X'}{\partial \eta} - \frac{\partial Y'}{\partial \xi}.$$

Obviously, the two systems of equations found for the system k must express exactly the same thing, since both systems of equations are equivalent to the Maxwell-Hertz equations for the system K. Further, since the equations for the two systems are in agreement apart from the symbols representing the vectors, it follows that the functions occurring in the systems of equations at corresponding places must agree up to a factor $\psi(v)$, common to all functions of one of the systems of equations and independent of ξ, η, ζ, and τ, but possibly depending on v. Thus we have the relations:

$$X' = \psi(v)X, \qquad L' = \psi(v)L,$$

$$Y' = \psi(v)\beta\left(Y - \frac{v}{V}N\right), \qquad M' = \psi(v)\beta\left(M + \frac{v}{V}Z\right),$$

$$Z' = \psi(v)\beta\left(Z + \frac{v}{V}M\right), \qquad N' = \psi(v)\beta\left(N - \frac{v}{V}Y\right).$$

If we now invert this system of equations, first by solving the equations just obtained, and second by applying to the equations the inverse transformation (from k to K), which is characterized by the velocity $-v$, we get, taking into account that both systems of equations so obtained must be identical,

$$\varphi(v) \cdot \varphi(-v) = 1.$$

Further, it follows for reasons of symmetry[4] that

$$\varphi(v) = \varphi(-v);$$

---

[4]If, e.g., $X = Y = Z = L = M = 0$ and $N \neq 0$, then it is clear for reasons of symmetry that if v changes its sign without changing its numerical value, then Y' must also change its sign without changing its numerical value.

ELECTRODYNAMICS OF MOVING BODIES

thus
$$\varphi(v) = 1,$$

and our equations take the form

$$X' = X, \qquad L' = L,$$
$$Y' = \beta\left(Y - \frac{v}{V}N\right), \qquad M' = \beta\left(M + \frac{v}{V}Z\right),$$
$$Z' = \beta\left(Z + \frac{v}{V}M\right), \qquad N' = \beta\left(N - \frac{v}{V}Y\right).$$

By way of interpreting these equations, we note the following remarks: Imagine a pointlike electric charge, whose magnitude measured in the rest system is "unit," i.e., which, when at rest in the rest system exerts a force of 1 dyne on an equal charge at a distance of 1 cm. According to the principle of relativity this electric charge is also of "unit" magnitude if measured in the moving system. If this electric charge is at rest relative to the rest system, then by definition the vector (X, Y, Z) equals the force acting on it. If, on the other hand, this acting charge is at rest relative to the moving system (at least at the relevant instant), then the force acting on it measured in the moving system is equal to the vector (X', Y', Z'). Hence, the first three of the above equations can be expressed in words in the following two ways:

1. If a unit point electric charge moves in an electromagnetic field, there acts upon it, in addition to the electric force, an "electromotive force" that, neglecting terms multiplied by the second and higher powers of v/V, is equal to the vector product of the velocity of the charge and the magnetic force, divided by the velocity of light. (Old mode of expression.)

2. If a unit point electric charge moves in an electromagnetic field, the force acting on it equals the electric force at the

145

location of the unit charge that is obtained by transforming the field to a coordinate system at rest relative to the unit charge. (New mode of expression.)

Analogous remarks hold for "magnetomotive forces."[8] We can see that in the theory developed here, the electromotive force only plays the role of an auxiliary concept, which owes its introduction to the circumstance that the electric and magnetic forces do not have an existence independent of the state of motion of the coordinate system.

It is further clear that the asymmetry in the treatment of currents produced by the relative motion of a magnet and a conductor, mentioned in the introduction, disappears. Moreover, questions about the "site" of electrodynamic electromotive forces (unipolar machines) become pointless.

7. Theory of Doppler's Principle and of Aberration

In the system K and very far from the coordinate origin, let there be a source of electrodynamic waves that, in a part of space containing the coordinate origin, are represented with sufficient accuracy by the equations

$$X = X_0 \sin \Phi, \quad L = L_0 \sin \Phi,$$

$$Y = Y_0 \sin \Phi, \quad M = M_0 \sin \Phi, \quad \Phi = \omega\left(t - \frac{ax + by + cz}{V} \right).$$

$$Z = Z_0 \sin \Phi, \quad N = N_0 \sin \Phi,$$

Here (X_0, Y_0, Z_0) and (L_0, M_0, N_0) are the vectors determining the amplitude of the wave train, and a, b, c are the direction cosines of the norm to the waves.

We want to know the character of these waves when investigated by an observer at rest in the moving system k.

ELECTRODYNAMICS OF MOVING BODIES

Applying the transformation equations for electric and magnetic forces found in section 6 and those for coordinates and
time found in section 3, we immediately obtain:

$$X' = X_0 \sin \Phi', \qquad L' = L_0 \sin \Phi',$$

$$Y' = \beta\left(Y_0 - \frac{v}{V}N_0\right)\sin \Phi', \qquad M' = \beta\left(M_0 + \frac{v}{V}Z_0\right)\sin \Phi',$$

$$Z' = \beta\left(Z_0 + \frac{v}{V}M_0\right)\sin \Phi', \qquad N' = \beta\left(N_0 - \frac{v}{V}Y_0\right)\sin \Phi',$$

$$\Phi' = \omega'\left(\tau - \frac{a'\xi + b'\eta + c'\zeta}{V}\right),$$

where we have put

$$\omega' = \omega\beta\left(1 - a\frac{v}{V}\right),$$

$$a' = \frac{a - \dfrac{v}{V}}{1 - a\dfrac{v}{V}},$$

$$b' = \frac{b}{\beta\left(1 - a\dfrac{v}{V}\right)},$$

$$c' = \frac{c}{\beta\left(1 - a\dfrac{v}{V}\right)}.$$

From the equation for ω' it follows that if an observer
moves with velocity v relative to an infinitely distant source
of light of frequency ν, in such a way that the connecting line
"light source–observer" forms an angle φ with the observer's
velocity, where this velocity is relative to a coordinate system
at rest relative to the light source, then ν', the frequency of

147

the light perceived by the observer, is given by the equation

$$\nu' = \nu \frac{1 - \cos \varphi \frac{v}{V}}{\sqrt{1 - \left(\frac{v}{V}\right)^2}}.$$

This is Doppler's principle for arbitrary velocities. For $\varphi = 0$ the equation takes the simple form

$$\nu' = \nu \sqrt{\frac{1 - \frac{v}{V}}{1 + \frac{v}{V}}}.$$

We see that, contrary to the usual conception, when $v = -\infty$, then $\nu = \infty$.[9]

If φ' denotes the angle between the wave normal (the direction of the ray) in the moving system and the connecting line "light source–observer,"[10] the equation for α'[11] takes the form

$$\cos \varphi' = \frac{\cos \varphi - \frac{v}{V}}{1 - \frac{v}{V} \cos \varphi}.$$

This equation expresses the law of aberration in its most general form. If $\varphi = \pi/2$, the equation takes the simple form

$$\cos \varphi' = -\frac{v}{V}.$$

We still need to find the amplitude of the waves as it appears in the moving system. If A and A' denote the amplitude of electric or magnetic force in the rest system and moving system respectively, we get

$$A'^2 = A^2 \frac{\left(1 - \frac{v}{V} \cos \varphi\right)^2}{1 - \left(\frac{v}{V}\right)^2},$$

which for $\varphi = 0$ takes the simpler form:

$$A'^2 = A^2 \frac{1 - \dfrac{v}{V}}{1 + \dfrac{v}{V}}.$$

It follows from these results that to an observer approaching a light source with velocity V, this source would have to appear infinitely intense.

8. Transformation of the Energy of Light Rays. Theory of Radiation Pressure Exerted on Perfect Mirrors

Since $A^2/8\pi$ equals the energy of light per unit volume, according to the principle of relativity we have to consider $A'^2/8\pi$ as the light energy in the moving system. Hence A'^2/A^2 would be the ratio of the energy of a given light complex "measured in motion" and "measured at rest" if the volume of a light complex were the same measured in K and in k. However, this is not the case. If a, b, c are the direction cosines of the wave normal of the light in the rest system, then the surface elements of the spherical surface

$$(x - Vat)^2 + (y - Vbt)^2 + (z - Vct)^2 = R^2$$

moving with the velocity of light are not traversed by any energy; we may therefore say that this surface permanently encloses the same light complex. We investigate the quantity of energy enclosed by this surface considered from the system k, i.e., the energy of the light complex relative to the system k.

Considered in the moving system, the spherical surface is an ellipsoidal surface whose equation at time $\tau = 0$ is

$$\left(\beta\xi - a\beta\frac{v}{V}\xi\right)^2 + \left(\eta - b\beta\frac{v}{V}\xi\right)^2 + \left(\zeta - c\beta\frac{v}{V}\xi\right)^2 = R^2.$$

If S denotes the volume of the sphere and S' that of the ellipsoid, then a simple calculation shows that

$$\frac{S'}{S} = \frac{\sqrt{1 - \left(\frac{v}{V}\right)^2}}{1 - \frac{v}{V}\cos\varphi}.$$

If we call the energy of the light enclosed by this surface E when measured in the rest system and E' when measured in the moving system, we obtain

$$\frac{E'}{E} = \frac{\frac{A'^2}{8\pi}S'}{\frac{A^2}{8\pi}S} = \frac{1 - \frac{v}{V}\cos\varphi}{\sqrt{1 - \left(\frac{v}{V}\right)^2}},$$

which, for $\varphi = 0$, simplifies to

$$\frac{E'}{E} = \sqrt{\frac{1 - \frac{v}{V}}{1 + \frac{v}{V}}}.$$

It is noteworthy that the energy and the frequency of a light complex vary with the observer's state of motion according to the same law.

Let the coordinate plane $\xi = 0$ be a completely reflecting surface, from which the plane waves considered in section 7 are reflected. We investigate the pressure of light exerted on the reflecting surface, and the direction, frequency, and intensity of the light after reflection.

Let the incident light be defined by the quantities A, $\cos \varphi$, and ν (relative to system K). Considered from k, the corresponding quantities are

$$A' = A \frac{1 - \frac{v}{V} \cos \varphi}{\sqrt{1 - \left(\frac{v}{V}\right)^2}},$$

$$\cos \varphi' = \frac{\cos \varphi - \frac{v}{V}}{1 - \frac{v}{V} \cos \varphi},$$

$$\nu' = \nu \frac{1 - \frac{v}{V} \cos \varphi}{\sqrt{1 - \left(\frac{v}{V}\right)^2}}.$$

Referring the process to the system k, we get for the reflected light

$$A'' = A',$$

$$\cos \varphi'' = - \cos \varphi',$$

$$\nu'' = \nu'.$$

Finally, by transforming back to the rest system K, we get for the reflected light

$$A''' = A'' \frac{1 + \frac{v}{V} \cos \varphi''}{\sqrt{1 - \left(\frac{v}{V}\right)^2}} = A \frac{1 - 2\frac{v}{V} \cos \varphi + \left(\frac{v}{V}\right)^2}{1 - \left(\frac{v}{V}\right)^2},$$

$$\cos \varphi''' = \frac{\cos \varphi'' + \frac{v}{V}}{1 + \frac{v}{V} \cos \varphi''} = - \frac{\left(1 + \left(\frac{v}{V}\right)^2\right) \cos \varphi - 2\frac{v}{V}}{1 - 2\frac{v}{V} \cos \varphi + \left(\frac{v}{V}\right)^2},$$

$$\nu''' = \nu'' \frac{1 + \frac{v}{V}\cos\varphi''}{\sqrt{1 - \left(\frac{v}{V}\right)^2}} = \nu \frac{1 - 2\frac{v}{V}\cos\varphi + \left(\frac{v}{V}\right)^2}{\left(1 - \frac{v}{V}\right)^2}. \quad \text{[12]}$$

The energy (measured in the rest system) that strikes a unit surface of the mirror per unit time is obviously $A^2/8\pi(V\cos\varphi - v)$. The energy leaving a unit surface of the mirror per unit time is $A'''^2/8\pi(-V\cos\varphi''' + v)$. According to the principle of energy conservation, the difference of these two expressions is the work done by light pressure per unit time. Equating this work to $P \cdot v$, where P is the pressure of light, we obtain

$$P = 2\,\frac{A^2}{8\pi}\,\frac{\left(\cos\varphi - \frac{v}{V}\right)^2}{1 - \left(\frac{v}{V}\right)^2}.$$

To first approximation, in agreement with experiment and with other theories, we get

$$P = 2\,\frac{A^2}{8\pi}\cos^2\varphi.$$

All problems in the optics of moving bodies can be solved by the method employed here. The essential point is that the electric and magnetic fields of light that is influenced by a moving body are transformed to a coordinate system that is at rest relative to that body. By this means, all problems in the optics of moving bodies are reduced to a series of problems in the optics of bodies at rest.

ELECTRODYNAMICS OF MOVING BODIES

9. Transformation of the Maxwell-Hertz Equations When Convection Currents Are Taken into Account

We start from the equations

$$\frac{1}{V}\left\{u_x\rho + \frac{\partial X}{\partial t}\right\} = \frac{\partial N}{\partial y} - \frac{\partial M}{\partial z}, \qquad \frac{1}{V}\frac{\partial L}{\partial t} = \frac{\partial Y}{\partial z} - \frac{\partial Z}{\partial y},$$

$$\frac{1}{V}\left\{u_y\rho + \frac{\partial Y}{\partial t}\right\} = \frac{\partial L}{\partial z} - \frac{\partial N}{\partial x}, \qquad \frac{1}{V}\frac{\partial M}{\partial t} = \frac{\partial Z}{\partial x} - \frac{\partial X}{\partial z},$$

$$\frac{1}{V}\left\{u_z\rho + \frac{\partial Z}{\partial t}\right\} = \frac{\partial M}{\partial x} - \frac{\partial L}{\partial y}, \qquad \frac{1}{V}\frac{\partial N}{\partial t} = \frac{\partial X}{\partial y} - \frac{\partial Y}{\partial x},$$

where

$$\rho = \frac{\partial X}{\partial x} + \frac{\partial Y}{\partial y} + \frac{\partial Z}{\partial z}$$

denotes 4π times the charge density, and (u_x, u_y, u_z) the velocity vector of the charge. If the electric charges are conceived as permanently bound to small, rigid bodies (ions, electrons), then these equations constitute the electromagnetic foundation of Lorentz's electrodynamics and optics of moving bodies.

If, using the transformation equations presented in sections 3 and 6, we transform these equations, assumed to be valid in the system K, to the system k, we get the equations

$$\frac{1}{V}\left\{u_\xi\rho' + \frac{\partial X'}{\partial \tau}\right\} = \frac{\partial N'}{\partial \eta} - \frac{\partial M'}{\partial \zeta}, \qquad \frac{1}{V}\frac{\partial L'}{\partial \tau} = \frac{\partial Y'}{\partial \zeta} - \frac{\partial Z'}{\partial \eta},$$

$$\frac{1}{V}\left\{u_\eta\rho' + \frac{\partial Y'}{\partial \tau}\right\} = \frac{\partial L'}{\partial \zeta} - \frac{\partial N'}{\partial \xi}, \qquad \frac{1}{V}\frac{\partial M'}{\partial \tau} = \frac{\partial Z'}{\partial \xi} - \frac{\partial X'}{\partial \zeta},$$

$$\frac{1}{V}\left\{u_\zeta\rho' + \frac{\partial Z'}{\partial \tau}\right\} = \frac{\partial M'}{\partial \xi} - \frac{\partial L'}{\partial \eta}, \qquad \frac{1}{V}\frac{\partial N'}{\partial \tau} = \frac{\partial X'}{\partial \eta} - \frac{\partial Y'}{\partial \xi},$$

153

where

$$\frac{u_x - v}{1 - \dfrac{u_x v}{V^2}} = u_\xi,$$

$$\frac{u_y}{\beta \left(1 - \dfrac{u_x v}{V^2} \right)} = u_\eta,$$

$$\frac{u_z}{\beta \left(1 - \dfrac{u_x v}{V^2} \right)} = u_\zeta,$$

and

$$\rho' = \frac{\partial X'}{\partial \xi} + \frac{\partial Y'}{\partial \eta} + \frac{\partial Z'}{\partial \zeta} = \beta \left(1 - \frac{v u_x}{V^2} \right) \rho.$$

Since—as follows from the velocity addition theorem (sec. 5)—the vector (u_ξ, u_η, u_ζ) is actually the velocity of the electric charges measured in the system k, we have thus shown that, on the basis of our kinematic principles, the electrodynamic foundation of Lorentz's theory of the electrodynamics of moving bodies agrees with the principle of relativity.

Let me also briefly add that the following important proposition can easily be deduced from the equations we have derived: If an electrically charged body moves arbitrarily in space without altering its charge when observed from a coordinate system moving with the body, then its charge also remains constant when observed from the "rest" system K.

10. Dynamics of the (Slowly Accelerated) Electron

In an electromagnetic field let there move an electrically charged particle with charge ϵ (called an "electron" in what follows); we assume only the following about its law of motion:

If the electron is at rest at a particular moment, its motion during the next instant of time will occur according to the equations

$$\mu\frac{d^2x}{dt^2} = \epsilon X,$$

$$\mu\frac{d^2y}{dt^2} = \epsilon Y,$$

$$\mu\frac{d^2z}{dt^2} = \epsilon Z,$$

where x, y, z denote the coordinates of the electron and μ its mass as long as the electron moves slowly.

Further, let the electron's velocity at a certain moment be v. We investigate the law of motion of the electron during the immediately succeeding instant of time.

Without loss of generality, we may and shall assume that the electron is at the coordinate origin and moves with velocity v along the X-axis of the system K at the moment with which we are concerned. It is then obvious that at the given moment ($t = 0$), the electron is at rest relative to a coordinate system k moving with constant velocity v parallel to the X-axis.

From the above assumption combined with the relativity principle, it is clear that, considered from the system k, the electron will move during the immediately ensuing period of time (for small values of t) according to the equations

$$\mu\frac{d^2\xi}{d\tau^2} = \epsilon X',$$

$$\mu\frac{d^2\eta}{d\tau^2} = \epsilon Y',$$

$$\mu\frac{d^2\zeta}{d\tau^2} = \epsilon Z',$$

where the symbols ξ, η, ζ, τ, X', Y', Z' all refer to the system k. If we also stipulate that, for $t = x = y = z = 0$, $\tau = \xi = \eta = \zeta = 0$ shall also hold, then the transformation equations of sections 3 and 6 are applicable, so that we get

$$\tau = \beta\left(t - \frac{v}{V^2}x\right),$$

$$\xi = \beta(x - vt), \qquad\qquad X' = x,$$

$$\eta = y, \qquad\qquad Y' = \beta\left(Y - \frac{v}{V}N\right),$$

$$\zeta = z, \qquad\qquad Z' = \beta\left(Z + \frac{v}{V}M\right).$$

With the help of these equations we transform the above equations of motion from the system k to the system K, obtaining

$$\frac{d^2x}{dt^2} = \frac{\epsilon}{\mu}\frac{1}{\beta^3}X,$$

$$\frac{d^2y}{dt^2} = \frac{\epsilon}{\mu}\frac{1}{\beta}\left(Y - \frac{v}{V}N\right), \qquad\qquad \text{(A)}$$

$$\frac{d^2z}{dt^2} = \frac{\epsilon}{\mu}\frac{1}{\beta}\left(Z + \frac{v}{V}M\right).$$

Following the usual approach, we now investigate the "longitudinal" and "transverse" mass of the moving electron. We write equations (A) in the form

$$\mu\beta^3\frac{d^2x}{dt^2} = \epsilon X = \epsilon X',$$

$$\mu\beta^2\frac{d^2y}{dt^2} = \epsilon\beta\left(Y - \frac{v}{V}N\right) = \epsilon Y',$$

$$\mu\beta^2\frac{d^2z}{dt^2} = \epsilon\beta\left(Z + \frac{v}{V}M\right) = \epsilon Z',$$

and note first that $\epsilon X'$, $\epsilon Y'$, $\epsilon Z'$ are the components of the ponderomotive force acting on the electron, as considered in a moving system that, at this instant, is moving with the same velocity as the electron. (This force could be measured, for example, by a spring balance at rest in the latter system.) If we simply call[13] this force "the force acting on the electron," and preserve the equation

$$\text{Mass} \times \text{Acceleration} = \text{Force},$$

stipulating, in addition, that accelerations be measured in the rest system K, then the above equations lead to the definition:

$$\text{Longitudinal mass} = \frac{\mu}{\left(\sqrt{1 - \left(\frac{v}{V}\right)^2}\right)^3},$$

$$\text{Transverse mass} = \frac{\mu}{1 - \left(\frac{v}{V}\right)^2}.$$

Of course, with a different definition of force and acceleration we would obtain different values for these masses; this shows that we must proceed very cautiously when comparing various theories of electron motion.

It should be noted that these results about mass are also valid for ponderable material points, because a ponderable material point can be made into an electron (in our sense of the word) by adding to it an *arbitrarily small* electric charge.

We now determine the kinetic energy of an electron. If an electron starts from the origin of the system K with an initial velocity 0 and continues to move along the X-axis under the influence of an electrostatic force X, it is clear that the energy it takes from the electrostatic field has the value

$\int \epsilon X dx$. Since the electron is supposed to accelerate slowly, and consequently cannot emit any energy in the form of radiation, the energy taken from the electrostatic field must be equated to the kinetic energy W of the electron. Bearing in mind that the first of equations (A) holds throughout the entire process of motion, we obtain

$$W = \int \epsilon X \, dx = \int_0^v \beta^3 v \, dv = \mu V^2 \left\{ \frac{1}{\sqrt{1 - \left(\frac{v}{V}\right)^2}} - 1 \right\}.$$

Thus, W becomes infinitely large when $v = V$. As is the case for our previous results, superluminal velocities are not possible.

By virtue of the argument presented above, this expression for kinetic energy must also be valid for ponderable masses.

Let us now enumerate the properties of the electron's motion resulting from the system of equations (A) that are accessible to experiment.

1. From the second equation of the system (A) it follows that an electric force Y and a magnetic force N have an equally strong deflective effect on an electron moving with velocity v if $Y = Nv/V$. Thus we see that, using our theory, it is possible to determine the velocity of the electron from the ratio of the magnetic deflection A_m to the electric deflection A_e for arbitrary velocities, by applying the law

$$\frac{A_m}{A_e} = \frac{v}{V}.$$

This relation can be tested experimentally since the velocity of the electron can also be measured directly, e.g., using rapidly oscillating electric and magnetic fields.

2. From the derivation of the kinetic energy of an electron it follows that the potential difference traversed by the electron and the velocity v that the electron acquires must be related by the equation

$$P = \int X \, dx = \frac{\mu}{\epsilon} V^2 \left\{ \frac{1}{\sqrt{1 - \left(\frac{v}{V}\right)^2}} - 1 \right\}.$$

3. We calculate the radius of curvature R of the path of the electron if a magnetic force N, acting perpendicularly to its velocity, is present (as the only deflecting force). From the second of equations (A) we obtain:

$$-\frac{d^2y}{dt^2} = \frac{v^2}{R} = \frac{\epsilon}{\mu} \frac{v}{V} N \cdot \sqrt{1 - \left(\frac{v}{V}\right)^2}$$

or

$$R = V^2 \frac{\mu}{\epsilon} \frac{\frac{v}{V}}{\sqrt{1 - \left(\frac{v}{V}\right)^2}} \cdot \frac{1}{N}.$$

These three relations are a complete expression of the laws by which, according to the theory presented here, the electron must move.

In conclusion, let me note that my friend and colleague M. Besso steadfastly stood by me in my work on the problem discussed here, and that I am indebted to him for several valuable suggestions.

(Annalen der Physik 17 [1905]: 891–921)

EDITORIAL NOTES

[1]In the 1913 reprint, a note was added after "be valid": "What is meant is, 'be valid in the first approximation.'" If Einstein did not write

PAPER 3

the additional notes to this paper, the contents of some of the notes suggest that he was consulted.

[2] Einstein introduces the designations Ξ, H, Z for the coordinates for the x', y', z' axes of the moving system.

[3] In the 1913 reprint, the following note is appended to the end of this line: "The Lorentz transformation equations are more simply derivable directly from the condition that, as a consequence of these equations, the relation $\xi^2 + \eta^2 + \zeta^2 - V^2\tau^2 = 0$ shall have the other $x^2 + y^2 + z^2 - V^2t^2 = 0$ as a consequence."

[4] See previous note.

[5] This result later became known as "the clock paradox." In 1911, Langevin seems to have first introduced human travelers, leading to the alternate name, "the twin paradox."

[6] In the 1913 reprint, the following note is appended to the word "Unruhuhr": "In contrast to the 'pendulum clock,' which—from the physical standpoint—is a system, to which the earth belongs; this had to be excluded."

[7] This fraction should be $\frac{w_\eta}{w_\xi}$.

[8] The term "motional magnetic force" was introduced by Heaviside. Einstein later defined the "magnetomotive force" as the force acting on a unit of magnetic charge moving through an electric field. To the order of approximation used in the discussion of "electromotive force," the magnetomotive force is given by $-1/V$ [$\mathbf{v}$, $\mathbf{E}$], where $\mathbf{E} = (L, M, N)$, $\mathbf{v} = (v, 0, 0)$, and the bracket is a vector product.

[9] Corrected by Einstein in a reprint copy to "when $v = -V$, then $\nu = \infty$."

[10] In ibid., "the connecting line 'light source–observer' " was canceled and interlineated with "direction of motion."

[11] α should be φ.

[12] In a reprint copy, the denominator in the final term is corrected to "$1 - (v/V)^2$."

[13] In the 1913 reprint, the following note is appended to "call": "The definition of force given here is not advantageous as was first noted by M. Planck. It is instead appropriate to define force in such a way that the laws of momentum and energy conservation take the simplest form."

160

PAPER 4

✦

Does the Inertia of a Body Depend on Its Energy Content?

THE RESULTS of an electrodynamic investigation recently published by me in this journal[1] lead to a very interesting conclusion, which will be derived here.

I based that investigation on the Maxwell-Hertz equations for empty space, together with Maxwell's expression for the electromagnetic energy of space, and also the following principle:

The laws according to which the states of physical systems change are independent of which one of the two coordinate systems (assumed to be in uniform parallel-translational motion relative to each other) is used to describe these changes (the principle of relativity).

Based on this foundation,[2] I derived the following result, among others (*loc. cit.*, sec. 8).

---

[1] A. Einstein, *Ann. d. Phys.* 17 (1905): 891. [See paper 3].

[2] The principle of the constancy of the velocity of light used there is of course contained in Maxwell's equations.

Let a system of plane light waves have the energy l relative to the coordinate system (x, y, z); let the ray direction (the wave-normal) make the angle φ with the x-axis of the system. If we introduce a new coordinate system (ξ, η, ζ), which is in uniform parallel translation with respect to the system (x, y, z), and the origin of which moves along the x-axis with velocity v, then this quantity of light—measured in the system (ξ, η, ζ)—has the energy

$$l^* = l \frac{1 - \dfrac{v}{V} \cos \varphi}{\sqrt{1 - \left(\dfrac{v}{V}\right)^2}},$$

where V denotes the velocity of light. We shall make use of this result in what follows.

Let there be a body at rest in the system (x, y, z) whose energy, relative to the system (x, y, z), is E_0. Let the energy of the body be H_0, relative to the system (ξ, η, ζ), moving with velocity v as above.

Let this body emit plane light waves of energy $L/2$ (measured relative to (x, y, z)) in a direction forming an angle φ with the x-axis, and at the same time an equal amount of light in the opposite direction. The body remains at rest with respect to system (x, y, z) during this process. This process must satisfy the principle of conservation of energy, and must be true (according to the principle of relativity) with respect to both coordinate systems. If E_1 and H_1 denote the energy of the body after emission of the light, measured relative to the system (x, y, z) and the (ξ, η, ζ), respectively, we obtain, using the relation indicated above,

$$E_0 = E_1 + \left[\frac{L}{2} + \frac{L}{2} \right],$$

$$H_0 = H_1 + \left[\frac{L}{2} \frac{1 - \frac{v}{V}\cos\varphi}{\sqrt{1 - \left(\frac{v}{V}\right)^2}} + \frac{L}{2} \frac{1 + \frac{v}{V}\cos\varphi}{\sqrt{1 - \left(\frac{v}{V}\right)^2}} \right]$$

$$= H_1 + \frac{L}{\sqrt{1 - \left(\frac{v}{V}\right)^2}}.$$

By subtraction, we obtain from these equations

$$(H_0 - E_0) - (H_1 - E_1) = L \left\{ \frac{1}{\sqrt{1 - \left(\frac{v}{V}\right)^2}} - 1 \right\}.$$

Both differences of the form $H - E$ occurring in this expression have simple physical meanings. H and E are the energy values of the same body, related to two coordinate systems in relative motion, the body being at rest in one of the systems (system (x, y, z)). Hence it is clear that the difference $H - E$ can differ from the body's kinetic energy K with respect to the other system (system (ξ, η, ζ)) only by an additive constant C, which depends on the choice of the arbitrary additive constants in the energies H and E. We can therefore set

$$H_0 - E_0 = K_0 + C$$
$$H_1 - E_1 = K_1 + C,$$

since C does not change during the emission of light. So we get

$$K_0 - K_1 = L \left\{ \frac{1}{\sqrt{1 - \left(\frac{v}{V}\right)^2}} - 1 \right\}.$$

The kinetic energy of the body with respect to (ξ, η, ζ) decreases as a result of emission of the light by an amount

PAPER 4

that is independent of the properties of the body. Furthermore, the difference $K_0 - K_1$ depends on the velocity in the same way as does the kinetic energy of an electron (*loc. cit.*, sec. 10).

Neglecting magnitudes of the fourth and higher order, we can get[1]

$$K_0 - K_1 = \frac{L}{V^2}\frac{v^2}{2}.$$

From this equation one immediately concludes:

If a body emits the energy L in the form of radiation, its mass decreases by L/V^2. Here it is obviously inessential that the energy taken from the body turns into radiant energy, so we are led to the more general conclusion:

The mass of a body is a measure of its energy content; if the energy changes by L, the mass changes in the same sense by $L/9 \cdot 10^{20}$ if the energy is measured in ergs and the mass in grams.

It is not excluded that it will prove possible to test this theory using bodies whose energy content is variable to a high degree (e.g., radium salts).

If the theory agrees with the facts, then radiation carries inertia between emitting and absorbing bodies.

(*Annalen der Physik* 18 [1905]: 639–641)

EDITORIAL NOTES

[1]Einstein used the Newtonian limit of the body's kinetic energy in order to evaluate change in its rest mass.

Part Four

✦

Einstein's Early Work on the
Quantum Hypothesis

Einstein in Bern, ca. 1905. (Lotte Jacobi
Archives, University of New Hampshire)

In describing his 1905 papers, Einstein characterized only paper 5 as revolutionary (see the Introduction, p. 5). It is still regarded as revolutionary for challenging the unlimited validity of Maxwell's theory of light and suggesting the existence of light quanta. Paper 5 shows that, at a sufficiently high frequency, the entropy of equilibrium thermal (or "black-body") radiation behaves as if the radiation consisted of a gas of independent "quanta of light energy" (or simply "light quanta"), each with energy proportional to the frequency of the corresponding wave. Einstein showed how to explain several otherwise puzzling phenomena by assuming that the interaction of light with matter consists of the emission or absorption of such light quanta.

Einstein became familiar with black-body radiation well before 1905. Mach's *Wärmelehre*, which Einstein read in 1897 or shortly thereafter, contains two chapters on thermal radiation, culminating in a discussion of Gustav Robert Kirchhoff's work. Kirchhoff showed that the energy emission spectrum of a perfectly black body (defined as one that absorbs all incident radiation) at a given temperature is a universal function of the temperature of the body and wavelength of the radiation. He inferred that equilibrium thermal radiation in a cavity with walls maintained at a certain temperature behaves like radiation emitted by a black body at the same temperature.

Heinrich Friedrich Weber, Einstein's physics professor at the Eidgenössische Technische Hochschule (ETH), was one of those who attempted to determine the universal black-body radiation function. He made measurements of the energy spectrum and proposed an empirical formula for the distribution function. He showed that, as a consequence of his formula, $\lambda_m = \text{constant}/T$ (where λ_m is the wavelength with the maximum intensity of the distribution), thus anticipating Wilhelm Wien's formulation of the displacement law for black-body radiation. Weber described his work in a course at the ETH given during the winter semester of 1898–1899, which Einstein took.

Einstein soon started to think seriously about the problem of radiation. By the spring of 1901, he was closely following Planck's work on black-body radiation. Originally, Planck had hoped to explain the irreversibility of physical processes by studying electromagnetic radiation; ultimately, he came to recognize that this could not be done without introducing statistical elements into the argument. In a series of papers between 1897 and 1900, Planck utilized Maxwell's electrodynamics to develop a theory of thermal radiation in interaction with one or more identical, charged harmonic oscillators within a cavity. He was only able to account for the radiation's irreversible approach to thermal equilibrium by employing methods analogous to those that Ludwig Boltzmann had used in kinetic theory. Planck introduced the notion of "natural" (that is, maximally disordered) radiation, which he defined by analogy with Boltzmann's definition of molecular chaos. Using Maxwell's theory, Planck derived a relation between the average energy $\bar{E}$ of a charged oscillator of frequency ν in equilibrium with thermal radiation and the energy density per unit frequency interval ρ_ν of the radiation at the same frequency:

$$\bar{E}_\nu = \frac{c^3}{8\pi\nu^2}\,\rho_\nu, \qquad (1)$$

where c is the speed of light.

Planck calculated the average energy of an oscillator by making assumptions about the entropy of the oscillators that enabled him to derive Wien's law for the energy density of the black-body spectrum, which originally seemed well supported by the experimental evidence. But by the turn of the century new observations showed systematic deviations from Wien's law for large values of λT.

Planck presented a new energy density distribution formula that agreed closely with observations over the entire spectrum:[1]

$$\rho_\nu = \frac{8\pi h \nu^3}{c^3} \frac{1}{e^{h\nu/kT} - 1}. \tag{2}$$

In this expression, now known as Planck's law or Planck's formula, $k = R/N$ is Boltzmann's constant, where R is the gas constant, N is Avogadro's (or Loschmidt's) number, and h is a new constant (later called Planck's constant). To derive this formula, Planck calculated the entropy of the oscillators, using what Einstein later called "the Boltzmann principle": $S = k \ln W$, where S is the entropy of a macroscopic state of the system, the probability of which is W. Following Boltzmann, Planck took the probability of a state to be proportional to the number of "complexions," or possible microconfigurations of the system corresponding to that state. He calculated this number by dividing the total energy of the state into a finite number of elements of equal magnitude, and counting the number of possible ways of distributing these energy elements among the individual oscillators. If the size of the energy elements is set equal to $h\nu$, where ν is the frequency of an oscillator, an expression for the entropy of an oscillator results that leads to eq. (2).

Although Einstein expressed private misgivings about Planck's approach in 1901, he did not mention Planck or black-body radiation in his papers until 1904. A study of the foundations of statistical physics, which he undertook between 1902 and 1904, provided Einstein with the tools he needed to analyze Planck's derivation and explore its consequences. At least three elements of Einstein's "general molecular theory of heat" were central to his subsequent work on the quantum hypothesis: (1) the introduction of the canonical ensemble; (2) the interpretation of probability as it occurs in Boltzmann's principle; and (3) the study of energy fluctuations in thermal equilibrium.

1. In an analysis of the canonical ensemble, Einstein proved that the equipartition theorem (see the Introduction, p. 16) holds for any system in thermal equilibrium. In paper 5 he showed that, when applied to an ensemble of charged harmonic oscillators in equilibrium with thermal radiation, the equipartition theorem leads, via eq. (1), to a black-body distribution law now known as the Rayleigh-Jeans law:

$$\rho_\nu = \frac{8\pi\nu^2}{c^3}\, kT. \tag{3}$$

Despite its rigorous foundation in classical physics, eq. (3) only agrees with the observed energy distribution for small values of ν/T; indeed, as Einstein noted, it implies an infinite total radiant energy.

2. In 1906, Einstein posed a question that preoccupied him and others at the time: "How is it that Planck did not arrive at the same formula [eq. (3)], but at the expression . . . [eq. (2)]?" One answer lies in Planck's definition of W in Boltzmann's principle, which, as Einstein repeatedly noted, differs fundamentally from his own definition of probabilities as time averages. As noted above, Planck interpreted W as

proportional to the number of complexions of a state of the system. As Einstein pointed out in 1909, such a definition of W is equivalent to its definition as the average, over a long period of time, of the fraction of time that the system spends in this state only if all complexions corresponding to a given total energy are equally probable. However, if this is assumed to hold for an ensemble of oscillators in thermal equilibrium with radiation, the Rayleigh-Jeans law results. Hence, the validity of Planck's law implies that all complexions cannot be equally probable. Einstein showed that, if the energies available to a canonical ensemble of oscillators are arbitrarily restricted to multiples of the energy element $h\nu$, then all possible complexions are not equally probable, and Planck's law results.

3. A third element of Einstein's work on statistical physics that is central to his work on the quantum hypothesis is his method for calculating mean-square fluctuations in the state variables of a system in thermal equilibrium. He employed the canonical ensemble to calculate such fluctuations in the energy of mechanical systems, then boldly applied the result to a nonmechanical system—black-body radiation, deducing a relation that agrees with Wien's displacement law discussed above. This agreement suggests the applicability of statistical concepts to radiation, and may have suggested to Einstein the possibility that radiation could be treated as a system with a finite number of degrees of freedom, a possibility he raised at the outset of paper 5.

In connection with his work on Brownian motion in 1905–1906, Einstein developed additional methods for calculating fluctuations, methods he later applied to the analysis of black-body radiation. In particular, he developed a method based on an inversion of Boltzmann's principle that can be used even in the absence of a microscopic model of the system. If the entropy of a system is given as a function of its macroscopic state variables, then Boltzmann's principle, in the form $W = \exp(S/k)$, can be used to calculate the probability of a state, and hence of fluctuations of any state variable. In 1909, Einstein used this method to calculate the fluctuations in the energy of black-body radiation in a given region of space. The stochastic method, used in the same paper to calculate fluctuations in the pressure of radiation, is based on his work on Brownian motion. Pressure fluctuations maintain the Brownian motion of a small mirror moving through the radiation field, in the face of the retarding force exerted on the mirror by the average radiation pressure. The results of these fluctuation calculations are discussed below.

Einstein's work on relativity also contributed to the development of his views on the nature of light. By eliminating the concept of the ether and showing that a flux of radiant energy transfers inertial mass, the theory of relativity demonstrated that light no longer need be treated as a disturbance in a hypothetical medium, but could be regarded as composed of independent structures, to which mass must be attributed.

AMONG Einstein's papers on the quantum hypothesis, paper 5 is unique in arguing for the notion of light quanta without using either the formal apparatus of his statistical papers or Planck's law. As noted above, Einstein demonstrated that only the Rayleigh-Jeans law, the limiting form of Planck's formula for small values of ν/T, is consistent with classical statistical mechanics and Maxwell's electrodynamics. At the other extreme, at which Wien's distribution law holds,

Einstein argues that "theoretical principles we have been using . . . fail completely." As he explained later that year, this failure "seems to me to have its basis in a fundamental incompleteness of our physical concepts."[2]

Einstein opened the paper by pointing out the "profound formal difference" between current theories of matter, in which the energy of a body is represented as a sum over a finite number of degrees of freedom, and Maxwell's theory, in which the energy is a continuous spatial function of fields having an infinite number of degrees of freedom. He suggested that the inability of Maxwell's theory to give an adequate account of radiation might be remedied by a theory in which radiant energy is distributed discontinuously in space. Einstein formulated "the light quantum hypothesis" that "when a light ray spreads out from a point source, the energy is not distributed continuously over an ever-increasing volume but consists of a finite number of energy quanta that are localized at points in space, move without dividing, and can be absorbed or generated only as complete units."

Using Wien's law, Einstein showed that the expression for the volume dependence of the entropy of radiation at a given frequency is similar in form to that of the entropy of an ideal gas. He concluded that "monochromatic radiation of low density (within the range of validity of Wien's radiation formula) behaves thermodynamically as if it consisted of mutually independent energy quanta of magnitude $R\beta v/N$."

In addition to its contributions to theory, paper 5 also provides ingenious explanations of several observed phenomena. It examines three interactions of light with matter, treated "as if light consisted of such energy quanta": Stokes's rule for fluorescence; the ionization of gases by ultraviolet light; and the photoelectric effect. Einstein proposed what later became known as his photoelectric equation,

$$E_{max} = (R/N)\beta\nu - P, \tag{4}$$

where E_{max} is the maximum kinetic energy of the photoelectrons, $R\beta/N$ is equivalent to Planck's h, ν is the frequency of the incident radiation, and P is the work function of the metal emitting the electrons. Although his derivation of this equation was later considered to be a leading achievement of that paper—it is cited in his 1922 Nobel Prize award—for almost two decades this argument failed to persuade most physicists of the validity of the light quantum hypothesis. Philip Lenard's experimental studies, to which Einstein referred, only provide qualitative evidence for an increase of E_{max} with frequency. For almost a decade the quantitative relationship between electron energy and radiation frequency was in doubt. By about 1914, there was a substantial body of evidence tending to support eq. (4). Robert Millikan's 1916 studies clinched the case for almost all physicists. But even the confirmation of Einstein's photoelectric equation did not bring about widespread acceptance of the concept of light quanta. Alternative interpretations of the photoelectric effect still received general support for a number of years.

The earliest widely accepted empirical evidence for the quantum hypothesis came not from radiation phenomena, but from data on specific heats of solids. In 1907, Einstein applied the quantum hypothesis to the model of a solid as a lattice of atoms harmonically bound to their equilibrium positions.[3] When the oscillators are treated classically, the equipartition theorem leads to the DuLong-Petit rule, predicting a constant specific heat for solids at all temperatures. Treating each atom as a quantized three-dimensional harmonic oscillator, Einstein was able to explain the well-known anomalous decrease of the specific heats of certain solids with decreasing temperature, and to obtain a relation between the specific heat of a solid and its selective absorption of infrared radiation.

Einstein had long suspected the existence of such a connection. Perhaps inspired by Planck's work, in 1901 Einstein wondered whether the internal kinetic energy of solids and liquids could be conceived of as "the energy of electric resonators." If it could, then the "specific heat and absorption spectrum of bodies must then be related."[4] He tried to connect this model with deviations from the DuLong-Petit rule.

In 1907 Einstein avoided the implications of the equipartition theorem for the specific heats of solids as he had avoided them for radiation theory, by introducing energy quanta. From the average energy of a quantized oscillator, Einstein derived an expression for the specific heat of a monatomic solid as a function of $\beta T/\nu$. The expression approaches zero with the temperature, and approaches the DuLong-Petit value at high temperatures. Considering the simplified nature of the model, the expression is in fairly good agreement with Weber's data on diamond.

A connection could also be made with absorption results from Drude's optical dispersion theory. Drude showed that the infrared optical eigenfrequencies of a solid are due to vibrations of the lattice ions, while the electrons are responsible for the ultraviolet eigenfrequencies. At room temperature, the value of Einstein's expression for specific heat effectively vanishes at a frequency well within the infrared region for most solids, and increases to the DuLong-Petit value for even lower frequencies. Einstein concluded that only the lattice ions and atoms contribute to the specific heats of solids. Moreover, if a solid displays infrared absorption resonances, the temperature dependence of its specific heat can be determined from these resonant frequencies.

In 1910 Nernst and his assistant Frederick A. Lindemann obtained general agreement between Einstein's predictions and observations of the variation with temperature of the specific heat of a number of solids. In 1911 Nernst reported the first confirmation of the quantum hypothesis outside the field of radiation, observing: "That the observations in their totality provide a brilliant confirmation of the quantum theory of Planck and Einstein is obvious."[5]

EDITORIAL NOTES

[1]Planck, *Annalen der Physik* 1 (1900): 719–737.

[2]Einstein, "Zur Theorie der Brownschen Bewegung," *Collected Papers*, vol. 2, doc. 32, pp. 334–345.

[3]"Planck's Theory of Radiation and the Theory of Specific Heats," *Annalen der Physik* 22 (1907): 180–190, reprinted in *Collected Papers*, vol. 2, doc. 38, pp. 379–389.

[4]Albert Einstein to Mileva Marić, 23 March 1901, *Collected Papers*, vol. 1, doc. 93.

[5]"Untersuchungen über die spezifische Wärme bei tiefen Temperaturen. III," *Königlich Preussische Akademie der Wissenschaften* (Berlin), *Sitzungsberichte* (1911), p. 310.

PAPER 5

✦

On a Heuristic Point of View
Concerning the Production and
Transformation of Light

A PROFOUND formal difference exists between the theoretical concepts that physicists have formed about gases and other ponderable bodies, and Maxwell's theory of electromagnetic processes in so-called empty space. While we consider the state of a body to be completely determined by the positions and velocities of an indeed very large yet finite number of atoms and electrons, we make use of continuous spatial functions to determine the electromagnetic state of a volume of space, so that a finite number of quantities cannot be considered as sufficient for the complete determination of the electromagnetic state of space. According to Maxwell's theory, energy is considered to be a continuous spatial function for all purely electromagnetic phenomena, hence also for light, whereas according to the present view of physicists, the energy of a ponderable body should be represented as a sum over the atoms and electrons. The energy of a ponderable body cannot be broken up into arbitrarily many, arbitrarily small parts, but according to Maxwell's theory (or, more generally, according to any wave theory) the energy of a light ray emitted from a point source continuously spreads out over an ever-increasing volume.

The wave theory of light, which operates with continuous spatial functions, has proved itself superbly in describing purely optical phenomena and will probably never be replaced by another theory. One should keep in mind, however, that optical observations refer to time averages rather than instantaneous values; and it is quite conceivable, despite the complete confirmation of the theory of diffraction, reflection, refraction, dispersion, etc., by experiment, that the theory of light, operating with continuous spatial functions, leads to contradictions when applied to the phenomena of emission and transformation of light.

Indeed, it seems to me that the observations of "blackbody radiation," photoluminescence, production of cathode rays by ultraviolet light, and other related phenomena associated with the emission or transformation of light appear more readily understood if one assumes that the energy of light is discontinuously distributed in space. According to the assumption considered here, in the propagation of a light ray emitted from a point source, the energy is not distributed continuously over ever-increasing volumes of space, but consists of a finite number of energy quanta localized at points of space that move without dividing, and can be absorbed or generated only as complete units.

In this paper I wish to present the train of thought and cite the facts that led me onto this path, in the hope that the approach to be presented will prove of use to some researchers in their investigations.

1. On a Difficulty Concerning the Theory of "Black-Body Radiation"

We shall begin by considering the following case from the perspective of Maxwell's theory and electron theory. Let a space enclosed by completely reflecting walls contain a number of freely moving gas molecules and electrons that exert conservative forces on each other when they come very close, i.e., can collide with each other like molecules according to the kinetic theory of gases.[1] Suppose, further, that a number of electrons are bound to widely separated points in space by forces directed toward these points and proportional to their distances from them. The bound electrons also enter into conservative interactions with the free molecules and electrons when the latter come very close. We call the bound electrons "resonators"; they emit and absorb electromagnetic waves of definite periods.

According to the present view concerning the origin of light, the radiation in the volume we are considering, as is found for the case of dynamic equilibrium based on Maxwell's theory, must be identical with "black-body radiation"—at least if one assumes that resonators of all relevant frequencies are present.

For the time being, we will disregard the radiation emitted and absorbed by the resonators and investigate the condition of dynamic equilibrium corresponding to the interac-

---

[1] This assumption is equivalent to the supposition that the mean kinetic energies of gas molecules and electrons are equal to each other at thermal equilibrium. As is well known, Mr. Drude used the latter assumption to derive a theoretical expression for the ratio of thermal and electrical conductivities of metals.

tion (collisions between) molecules and electrons. For such an equilibrium, the kinetic theory of gases asserts that the mean kinetic energy of a resonator electron must be equal to the mean translational kinetic energy of a gas molecule. If we decompose the motion of a resonator electron into three mutually perpendicular oscillatory motions, we find for the mean value $\bar{E}$ of the energy of one such linear oscillatory motion

$$\bar{E} = \frac{R}{N}T,$$

where R denotes the universal gas constant, N the number of "real molecules"[1] in one gram-equivalent, and T the absolute temperature. Because of the equality of the time averages of the resonator's kinetic and potential energies, the energy $\bar{E}$ is two-thirds the kinetic energy of a free monatomic gas molecule. If due to some cause—in our case, radiation processes—the energy of a resonator were to have a time average greater or less than $\bar{E}$, then collisions with the free electrons and molecules would lead to an energy loss to or a gain from the gas, different on the average from zero. Thus, in the case we are considering, dynamic equilibrium is possible only if the average energy of each resonator is $\bar{E}$.

We will now apply a similar argument to the interaction between the resonators and the radiation present in space. Mr. Planck has derived the condition for dynamic equilibrium in this case[2] on the assumption that the radiation can be treated as the most completely conceivable disordered

[2] M. Planck, *Ann. d. Phys.* 1 (1900): 99.

process.[3] He found:

$$\bar{E}_\nu = \frac{L^3}{8\pi\nu^2}\rho_\nu.$$

$\bar{E}_\nu$ is here the mean energy of a resonator with eigenfrequency ν (per unit frequency interval), L is the velocity of light, ν the frequency, and $\rho_\nu \, d\nu$ the energy per unit volume of that part of the radiation whose frequency lies between ν and $\nu + d\nu$.

[3]This assumption can be formulated as follows. We expand the Z-component of the electrical force (Z) at an arbitrary point of the space being considered during the time interval between $t = 0$ and $t = T$ (where T denotes a time period very large relative to all relevant oscillation periods) in a Fourier series,

$$Z = \sum_{\nu=1}^{\nu=\infty} A_\nu \sin\left(2\pi\nu\frac{t}{T} + \alpha_\nu\right),$$

where $A_\nu \geq 0$ and $0 \leq \alpha_\nu \leq 2\pi$. If one imagines that at the same point of space such an expansion is made arbitrarily often at randomly chosen starting times, then one will obtain different sets of values for the quantities A_ν and α_ν. For the frequency of occurrence of the various combinations of values of the quantities A_ν and α_ν, there will then exist (statistical) probabilities dW of the form

$$dW = f(A_1 A_2 \ldots \alpha_1\alpha_2 \ldots) \, dA_1 dA_2 \ldots d\alpha_1 d\alpha_2 \ldots$$

The radiation is then most conceivably disordered if

$$f(A_1, A_2 \ldots \alpha_1, \alpha_2 \ldots) = F_1(A_1)F_2(A_2) \ldots f_1(\alpha_1) \cdot f_2(\alpha_2) \ldots,$$

i.e., when the probability of a specific value of one of the quantities A or x[2] is independent of the values taken by the other quantities A and α respectively. Hence, the more closely fulfilled the condition is that individual pairs of quantities A_ν and α_ν depend on the emission and absorption processes of particular groups of resonators, the more closely can radiation in our case be considered to be "most conceivably disordered."

If, on the whole, the radiation energy of frequency ν does not continually decrease or increase, the following relations must hold:

$$\frac{R}{N}T = \bar{E} = \bar{E}_\nu = \frac{L^3}{8\pi\nu^2}\rho_\nu,$$

$$\rho_\nu = \frac{R}{N}\frac{8\pi\nu^2}{L^3}T. \text{[3]}$$

These relations, found as the conditions of dynamic equilibrium, not only fail to agree with experiment but also imply that in our model a definite distribution of energy between ether and matter is out of the question. Indeed, the wider the range of the resonator frequencies chosen, the greater the total radiation energy of the volume, and in the limit we obtain

$$\int_0^\infty \rho_\nu \, d\nu = \frac{R}{N}\frac{8\pi}{L^3}T\int_0^\infty \nu^2 \, d\nu = \infty.$$

2. ON PLANCK'S DETERMINATION OF THE ELEMENTARY QUANTA[4]

We now wish to show that Mr. Planck's determination of the elementary quanta is to a certain extent independent of his theory of "black-body radiation."

Planck's formula[4] for ρ_ν, which is satisfied by all experiments up to now, reads

$$\rho_\nu = \frac{\alpha\nu^3}{e^{\frac{\beta\nu}{T}} - 1},$$

[4] M. Planck, Ann. d. Phys. 4 (1901): 561.

where

$$\alpha = 6.10 \times 10^{-56}$$
$$\beta = 4.866 \times 10^{-11}.$$

For large values of T/ν, i.e., for high wavelengths and radiation densities, this equation takes the following limiting form:

$$\rho_\nu = \frac{\alpha}{\beta} \nu^2 T.$$

One can see that this formula agrees with that derived in section 1 from Maxwell's theory and electron theory. By equating the coefficients of the two formulas, we obtain

$$\frac{R}{N} \frac{8\pi}{L^3} = \frac{\alpha}{\beta}$$

or

$$N = \frac{\beta}{\alpha} \frac{8\pi R}{L^3} = 6.17 \times 10^{23},$$

i.e., one atom of hydrogen weighs $1/N$ gram $= 1.62 \times 10^{-24}$ g. This is exactly the value found by Mr. Planck, which in turn is in satisfactory agreement with the values found by other methods.

We therefore arrive at the following conclusion: the higher the energy density and wavelength of radiation, the more reasonable the theoretical foundations we have been using prove to be; however, they fail completely in the case of low wavelengths and low radiation densities.

In the following, we shall consider "black-body radiation" together with the experimental facts, without establishing any model for the emission and propagation of radiation.

3. On the Entropy of Radiation

The following treatment is contained in a well-known study by Mr. Wien and is presented here only for the sake of completeness.

Consider radiation occupying a volume v. We assume that the observable properties of this radiation are completely determined when the radiation density $\rho(\nu)$ is given for all frequencies.[5] Since radiations of different frequencies can be regarded as separable from each other without performing any work or transferring any heat, the entropy of the radiation can be represented by

$$S = v \int_0^\infty \varphi(\rho, \nu) \, d\nu,$$

where φ is a function of the variables ρ and ν. One can reduce φ to a function of a single variable by asserting that adiabatic compression of radiation between reflecting walls does not change its entropy. However, we shall not pursue this, but will immediately investigate how the function φ can be obtained from the black-body radiation law.

In the case of "black-body radiation," ρ is such a function of ν that the entropy is a maximum for a given energy, i.e.,

$$\delta \int_0^\infty \varphi(\rho, \nu) \, d\nu = 0$$

if

$$\delta \int_0^\infty \rho \, d\nu = 0.$$

---

[5] This assumption is arbitrary. We shall naturally keep this simplest assumption as long as experiment does not force us to abandon it.

From this it follows that for every choice of $\delta\rho$ as a function of ν,

$$\int_0^\infty \left(\frac{\partial\varphi}{\partial\rho} - \lambda \right) \delta\rho \, d\nu = 0,$$

where λ is independent of ν. Thus, for black-body radiation, $\partial\varphi/\partial\rho$ is independent of ν.

The following equation applies when the temperature of black-body radiation of volume $v = 1$ increases by dT:

$$dS = \int_{\nu=0}^{\nu=\infty} \frac{\partial\varphi}{\partial\rho} \, d\rho \, d\nu,$$

or, since $\partial\varphi/\partial\rho$ is independent of ν,

$$dS = \frac{\partial\varphi}{\partial\rho} \, dE.$$

Since dE is equal to the heat added and the process is reversible, we also have

$$dS = \frac{1}{T} \, dE.$$

Comparison yields

$$\frac{\partial\varphi}{\partial\rho} = \frac{1}{T}.$$

This is the law of black-body radiation. Thus, one can derive the law of black-body radiation from the function φ, and, inversely, the function φ can be determined by integration, remembering that φ vanishes for $\rho = 0$.

4. Limiting Law for the Entropy of Monochromatic Radiation at Low Radiation Density

The existing observations of "black-body radiation" show that the law

$$\rho = \alpha \nu^3 e^{-\beta \frac{\nu}{T}}$$

originally postulated by Mr. W. Wien for "black-body radiation" is not strictly valid. However, it has been fully confirmed by experiment for large values of ν/T. We shall base our calculations on this formula, bearing in mind, however, that our results are valid only within certain limits.

First of all, this formula yields

$$\frac{1}{T} = -\frac{1}{\beta \nu} \ln \frac{\rho}{\alpha \nu^3},$$

and next, using the relation obtained in the preceding section,

$$\varphi(\rho, \nu) = -\frac{\rho}{\beta \nu} \left\{ \ln \frac{\rho}{\alpha \nu^3} - 1 \right\}.$$

Suppose now that we have radiation of energy E, with frequency between ν and $\nu + d\nu$, occupying a volume v. The entropy of this radiation is

$$S = v \varphi(\rho, \nu) \, d\nu = -\frac{E}{\beta \nu} \left\{ \ln \frac{E}{v \alpha \nu^3 d\nu} - 1 \right\}. \text{[5]}$$

If we restrict ourselves to investigating the dependence of the entropy on the volume occupied by the radiation, and denote the entropy of radiation by S_0 at volume v_0, we obtain

$$S - S_0 = \frac{E}{\beta \nu} \ln \left[\frac{v}{v_0} \right].$$

This equation shows that the entropy of monochromatic radiation of sufficiently low density varies with the volume according to the same law as the entropy of an ideal gas or a dilute solution. In the following, we shall interpret this equation on the basis of the principle introduced into physics by Mr. Boltzmann, according to which the entropy of a system is a function of the probability of its state.

5. MOLECULAR-THEORETICAL INVESTIGATION OF THE DEPENDENCE ON VOLUME OF THE ENTROPY OF GASES AND DILUTE SOLUTIONS

In calculating entropy by molecular-theoretical methods, the word "probability" is often used in a sense differing from the way the word is defined in probability theory. In particular, "cases of equal probability" are often hypothetically stipulated when the theoretical models employed are definite enough to permit a deduction rather than a stipulation. I will show in a separate paper that, in dealing with thermal processes, it suffices completely to use the so-called statistical probability, and hope thereby to eliminate a logical difficulty that still obstructs the application of Boltzmann's principle. Here, however, I shall give only its general formulation and its application to very special cases.

If it is meaningful to talk about the probability of a state of a system, and if, furthermore, each increase of entropy can be conceived as a transition to a state of higher probability, then the entropy S_1 of a system is a function of the probability W_1 of its instantaneous state. Therefore, if we have two systems S_1 and S_2 that do not interact with each other, we can set:

$$S_1 = \varphi_1(W_1),$$
$$S_2 = \varphi_2(W_2).$$

If these two systems are viewed as a single system of entropy S and probability W, then

$$S = S_1 + S_2 = \varphi(W)$$

and

$$W = W_1 \cdot W_2.$$

The last equation tells us that the states of the two systems are events that are independent of each other.

From these equations it follows that

$$\varphi(W_1 \cdot W_2) = \varphi_1(W_1) + \varphi_2(W_2),$$

and, finally,

$$\varphi_1(W_1) = C \ln(W_1) + \text{const.}$$
$$\varphi_2(W_2) = C \ln(W_2) + \text{const.}$$
$$\varphi(W) = C \ln(W) + \text{const.}$$

The quantity C is therefore a universal constant; it follows from the kinetic theory of gases that its value is R/N, where the meaning of the constants R and N is the same as above. If S_0 denotes the entropy of a certain initial state and W is the relative probability of a state of entropy S, we obtain in general:

$$S - S_0 = \frac{R}{N} \ln W.$$

We now treat the following special case. Let a volume v_0 contain a number (n) of moving points (e.g., molecules) to which our discussion will apply. The volume may also contain any arbitrary number of other moving points of any kind. No assumption is made about the law governing the motion of the points under discussion in the volume except that, as concerns this motion, no part of the space (and no direction within it) is preferred over the others. Further, let the number of (aforementioned) moving points under discussion be small enough that we can disregard interactions between them.

This system, which, for example, can be an ideal gas or a dilute solution, possesses a certain entropy S_0. Let us imagine that all n moving points are assembled in a part of the volume v_0 of size v without any other changes in the system. It is obvious that this state has a different value of entropy (S), and we now wish to determine the difference in entropy with the help of Boltzmann's principle.

We ask: How great is the probability of the last-mentioned state relative to the original one? Or: How great is the probability that at a randomly chosen instant of time, all n independently moving points in a given volume v_0 will be found (by chance) in the volume v?

Obviously, this probability, which is a "statistical probability," has the value

$$W = \left(\frac{v}{v_0}\right)^n;$$

from this, by applying Boltzmann's principle, one obtains

$$S - S_0 = R\left(\frac{n}{N}\right)\ln\left(\frac{v}{v_0}\right).$$

It is noteworthy that, in the derivation of this equation, from which the Boyle–Gay-Lussac law and the analogous law of osmotic pressure can easily be derived thermo-dynamically,[6] no assumptions need be made about the law governing the motion of the molecules.

6. Interpretation According to Boltzmann's Principle of the Expression for the Dependence of the Entropy of Monochromatic Radiation on Volume

In section 4 we found the following expression for the dependence of the entropy of monochromatic radiation on volume:

$$S - S_0 = \frac{E}{\beta\nu} \ln\left(\frac{v}{v_0}\right).$$

If we write this formula in the form

$$S - S_0 = \frac{R}{N} \ln\left[\left(\frac{v}{v_0}\right)^{\frac{N}{R}\frac{E}{\beta\nu}}\right]$$

and compare it with the general formula expressing the Boltzmann principle,

$$S - S_0 = \frac{R}{N} \ln W,$$

we arrive at the following conclusion: If monochromatic radiation of frequency ν and energy E is enclosed (by reflecting

---

[6] If E is the energy of the system, we get

$$-d(E - TS) = p\,dv = T\,dS = R\frac{n}{N}\frac{dv}{v};[6]$$

thus

$$pv = R\,\frac{n}{N}T.$$

walls) in the volume v_0, the probability that at a randomly chosen instant the total radiation energy will be found in the portion v of the volume v_0 is

$$W = \left(\frac{v}{v_0}\right)^{\frac{N}{R}\frac{E}{\beta\nu}}.$$

From this we further conclude that monochromatic radiation of low density (within the range of validity of Wien's radiation formula) behaves thermodynamically as if it consisted of mutually independent energy quanta of magnitude $R\beta\nu/N$.[7]

We also want to compare the mean value of the energy quanta of black-body radiation with the mean kinetic energy of the center-of-mass motion of a molecule at the same temperature. The latter is $\frac{3}{2}(R/N)T$, while the mean value of the energy quantum obtained on the basis of Wien's formula is

$$\frac{\int_0^\infty \alpha\nu^3 e^{-\frac{\beta\nu}{T}}\, d\nu}{\int_0^\infty \frac{N}{R\beta\nu}\alpha\nu^3 e^{-\frac{\beta\nu}{T}}\, d\nu} = 3\frac{R}{N}T.$$

If monochromatic radiation (of sufficiently low density) behaves, as concerns the dependence of its entropy on volume, as though the radiation were a discontinuous medium consisting of energy quanta of magnitude $R\beta\nu/N$, then it seems reasonable to investigate whether the laws governing the emission and transformation of light are also constructed as if light consisted of such energy quanta. We will consider this question in the following sections.

191

7. On Stokes's Rule

Let monochromatic light be transformed by photoluminescence into light of a different frequency, and, in accordance with the result just obtained, let us assume that both the incident and emitted light consist of energy quanta of magnitude $(R/N)\beta\nu$, where ν denotes the relevant frequency. The transformation process can then be interpreted as follows. Each incident energy quantum of frequency ν_1 is absorbed and—at least at a sufficiently low distribution density of the incident energy quanta—generates by itself a light quantum of frequency ν_2; it is possible that the absorption of the incident light quantum may simultaneously give rise to the emission of light quanta of frequencies ν_3, ν_4, etc., as well as to energy of some other kind (e.g., heat). It makes no difference by means of what intermediary processes this end result occurs. If the photoluminescent substance is not regarded as a permanent energy source, then, according to the principle of conservation of energy, the energy of an emitted energy quantum cannot be greater than that of the light quantum that produced it; hence it follows that

$$\frac{R}{N}\beta\nu_2 \leqq \frac{R}{N}\beta\nu_1,$$

or

$$\nu_2 \leqq \nu_1.$$

This is the well-known Stokes's rule.

It is to be emphasized that, according to our conception, at low illumination the amount of light emitted must be proportional to the intensity of the incident light, because each incident energy quantum will induce an elementary process of the kind indicated above, independently of other incident energy quanta. In particular, there will be no lower limit for the intensity of the incident light, below which the light would be unable to excite the fluorescent effect. According to the conception of the phenomenon presented here, deviations from Stokes's rule are conceivable in the following cases:

1. If the number of energy quanta per unit volume simultaneously being transformed is so large that an energy quantum of emitted light can obtain its energy from several incident energy quanta.

2. When the incident (or emitted) light does not have the same energy distribution as black-body radiation within the range of validity of Wien's law; if, for example, the incident light is produced by a body of such high temperature that Wien's law is no longer valid for the relevant wavelengths.

The latter possibility deserves special attention. According to the conception developed above, it is indeed not impossible that even at very low densities a "non-Wien radiation" will behave differently as concerns its energy from black-body radiation within the range of validity of Wien's law.

8. ON THE GENERATION OF CATHODE RAYS BY ILLUMINATION OF SOLID BODIES

The usual view that the energy of light is continuously distributed over the space through which it travels faces especially great difficulties when one attempts to explain photoelectric phenomena, which are expounded in a pioneering work by Mr. Lenard.[7]

[7] P. Lenard, *Ann. d. Phys.* 8 (1902): 169, 170.

According to the view that the incident light consists of energy quanta of energy $(R/N)\beta\nu$, the production of cathode rays by light can be conceived in the following way. The body's surface layer is penetrated by energy quanta whose energy is converted at least partially into kinetic energy of the electrons. The simplest conception is that a light quantum transfers its entire energy to a single electron; we will assume that this can occur. However, we will not exclude the possibility that electrons absorb only a part of the energy of the light quanta.

An electron in the interior of the body having kinetic energy will have lost part of its kinetic energy by the time it reaches the surface. In addition, we will assume that, in leaving the body, each electron must perform some work, P (characteristic for the body). The electrons leaving the body with the greatest perpendicular velocity will be those located right on the surface and ejected normally to it. The kinetic energy of such electrons is

$$\frac{R}{N}\beta\nu - P.$$

If the body is charged to a positive potential Π and surrounded by conductors at zero potential, and if Π is just sufficient to prevent a loss of electrical charge by the body, it follows that

$$\Pi\epsilon = \frac{R}{N}\beta\nu - P,$$

where ϵ denotes the charge of the electron; or

$$\Pi E = R\beta\nu - P',$$

where E is the charge of one gram-equivalent of a monovalent ion and P' is the potential of this quantity of negative charge relative to the body.[8]

If one sets $E = 9.6 \times 10^3$, then $\Pi \cdot 10^{-8}$ is the potential in volts that the body acquires when irradiated in vacuum.

To see whether the derived relation agrees in order of magnitude with experiment, set $P' = 0$, $\nu = 1.03 \times 10^{15}$ (which corresponds to the ultraviolet limit of the solar spectrum) and $\beta = 4.866 \times 10^{-11}$. We obtain $\Pi \cdot 10^7 = 4.3$ volts, a result that agrees in order of magnitude with the results of Mr. Lenard.[9]

If the formula derived is correct, then Π, when plotted in Cartesian coordinates as a function of the frequency of the incident light, must give a straight line whose slope is independent of the nature of the substance under study.

As far as I can tell, this conception of the photoelectric effect does contradict its properties as observed by Mr. Lenard. If each energy quantum of the incident light transmits its energy to electrons, independently of all others, then the velocity distribution of the electrons, i.e., the nature of cathode rays produced, will be independent of the intensity of the incident light; on the other hand, under otherwise identical circumstances, the number of electrons leaving the body will be proportional to the intensity of the incident light.[10]

---

[8] If one assumes that the individual electron can only be detached by light from a neutral molecule by the expenditure of some work, one does not have to change anything in the relation just derived; one need only consider P' as the sum of two terms.

[9] P. Lenard, *Ann. d. Phys.* 8 (1902): 165, 184, table I, fig. 2.

[10] P. Lenard, *loc. cit.*, pp. 150, 166–168.

Remarks similar to those regarding possible deviations from Stokes's rule apply to the possible limits of validity of the laws set forth above.

In the foregoing it has been assumed that at least some of the quanta of the incident light transmit their energies completely to one electron each. If this likely assumption is not made, one obtains the following equation in place of the last one:

$$\Pi E + P' \leqq R\beta\nu.$$

For cathode luminescence, which is the inverse of the process discussed above, one obtains by analogous considerations:

$$\Pi E + P' \geqq R \beta\nu.$$

For the substances investigated by Mr. Lenard, $PE^{[8]}$ is always considerably greater than $R\beta\nu$ because the potential difference, which the cathode rays must traverse in order to produce visible light, amounts in some cases to hundreds, in others to thousands of volts.[11] We must therefore assume that the kinetic energy of one electron goes into the production of many energy quanta of light.

9. On the Ionization of Gases by Ultraviolet Light

We have to assume that, in ionization of a gas by ultraviolet light, one energy quantum of light serves to ionize one gas molecule. From this it follows that the ionization energy (i.e., the work theoretically required for its ionization) of a

---

[11] P. Lenard, *Ann. d. Phys.* 12 (1903): 469

molecule cannot be greater than the energy of an absorbed light quantum capable of producing this effect. If J denotes the (theoretical) ionization energy per gram-equivalent, it follows that

$$R\beta\nu \geqq J.$$

According to Lenard's measurements, the largest effective wavelength for air is about 1.9×10^{-5} cm, hence

$$R\beta\nu = 6.4 \times 10^{12} \text{ erg} \geqq J.$$

An upper limit for the ionization energy can also be obtained from the ionization potentials of rarefied gases. According to J. Stark[12] the smallest measured ionization potential (for platinum anodes) for air is about 10 volts.[13] Thus, one obtains 9.6×10^{12} as the upper limit for J, which is almost equal to the value found above. There is still another consequence, verification of which by experiment seems to me of great importance. If each absorbed energy quantum of light ionizes one molecule, then the following relation must hold between the quantity of light absorbed L and the number j of gram-molecules of ionized gas:

$$j = \frac{L}{R\beta\nu}.$$

If our conception is correct, this relation must be valid for all gases that (at the relevant frequency) show no appreciable absorption unaccompanied by ionization.

(*Annalen der Physik* 17 [1905]: 132–148)

---

[12] J. Stark, *Die Elektrizität in Gasen*, p. 57 (Leipzig, 1902).
[13] In the interior of the gases, however, the ionization potential of negative ions is five times greater.

EDITORIAL NOTES

[1] "Real molecules" are presumably those that are not dissociated.

[2] x should be α.

[3] Expressions equivalent to Einstein's equation were obtained in 1905 by Rayleigh and Jeans without the use of material resonators.

[4] Here "elementary quanta" refers to fundamental atomic constants. In 1901, Planck determined the mass of the hydrogen atom, Loschmidt's number (N), Boltzmann's constant, and the elementary electric charge.

[5] S refers to radiation with frequencies between ν and $\nu + d\nu$, and $E = \rho\nu\,d\nu$.

[6] The last term should be multiplied by T.

[7] $R\beta/N$ corresponds to Planck's "h."

[8] PE should be ΠE.